SpringerBriefs in Ethics

Springer Briefs in Ethics envisions a series of short publications in areas such as business ethics, bioethics, science and engineering ethics, food and agricultural ethics, environmental ethics, human rights and the like. The intention is to present concise summaries of cutting-edge research and practical applications across a wide spectrum.

Springer Briefs in Ethics are seen as complementing monographs and journal articles with compact volumes of 50 to 125 pages, covering a wide range of content from professional to academic. Typical topics might include:

- Timely reports on state-of-the art analytical techniques
- A bridge between new research results, as published in journal articles, and a contextual literature review
- A snapshot of a hot or emerging topic
- In-depth case studies or clinical examples
- Presentations of core concepts that students must understand in order to make independent contributions

More information about this series at http://www.springer.com/series/10184

Euzebiusz Jamrozik · Michael J. Selgelid

Human Challenge Studies in Endemic Settings

Ethical and Regulatory Issues

 Springer Open

Euzebiusz Jamrozik
Monash Bioethics Centre
Monash University
Melbourne, VIC, Australia

Department of Medicine
Royal Melbourne Hospital
University of Melbourne
Melbourne, VIC, Australia

Nuffield Department of Population Health
Wellcome Centre for Ethics and the
Humanities and The Ethox Centre
University of Oxford
Oxford, UK

Michael J. Selgelid
Monash Bioethics Centre
Monash University
Clayton, VIC, Australia

ISSN 2211-8101   ISSN 2211-811X (electronic)
SpringerBriefs in Ethics
ISBN 978-3-030-41479-5   ISBN 978-3-030-41480-1   (eBook)
https://doi.org/10.1007/978-3-030-41480-1

This Springer imprint is published by the registered company Springer Nature Switzerland AG
The registered company address is: Gewerbestrasse 11, 6330 Cham, Switzerland

Preface

This volume is comprised of the Final Report of a project commissioned by the Wellcome Trust. In response to its request for a "comparative review of existing controlled human infection studies in endemic settings to inform the development of an international ethics framework", this Report aims to (1) outline the primary ethical and regulatory issues associated with human challenge studies in general, (2) examine the implications of such issues (or any special/unique issues arising) in endemic regions, and (3) provide detailed case studies of endemic region human challenge studies. It is informed by reviews of relevant scientific and ethical literature, as well as in-depth interviews with science and ethics experts.

Clayton, Australia

Euzebiusz Jamrozik
Michael J. Selgelid

Acknowledgments

This work was supported by the Wellcome Trust [210551/Z/18/Z] with Michael Selgelid as primary investigator and Euzebiusz Jamrozik as lead researcher. We thank Jane Brophy for conducting some of the interviews for this project. Euzebiusz Jamrozik is grateful to the Brocher Foundation for providing the opportunity to work on this project while a Brocher scholar in residence. We would also like to extend our gratitude to all expert stakeholders interviewed, and to the staff of the Wellcome Trust for their support throughout this project.

Executive Summary

Background

Human infection challenge studies (HCS) are experiments that involve infecting research participants with pathogens (or other micro-organisms) and primarily aim to (i) test (novel) vaccines and/or therapeutics, (ii) generate knowledge regarding the natural history of infectious diseases (and/or asymptomatic infection), and/or (iii) develop "models of infection"—i.e., reliable methods (to be used in studies with aims (i) and/or (ii)) of infecting human research participants with particular pathogens. Modern HCS are sometimes referred to as "controlled human infection studies," because they involve *controlling* the selection and/or production of the microorganism strain (s) and the timing, route, and/or dose of infection; infection in a *controlled* environment; and/or (with the aim to avoid serious harm to research participants) infection with microorganisms causing no disease or disease that is self-limiting or can be (and is) *controlled* with effective cures or treatments; and/or *controlling* who is being infected (and/or subjected to other experimental interventions).

HCS have a good safety record over the last 50 years as a result of careful research practices, including close monitoring of participants. They are ethically sensitive, however, and raise complex questions regarding, *inter alia*, (i) the acceptable limit of risks and other burdens which may be imposed on healthy volunteers (ii) the need for protection of third parties from infection (by participants), (iii) appropriate criteria for participant selection/exclusion (e.g., determinations of when, if ever, it would be justifiable to recruit participants from especially vulnerable populations, including children), (iv) appropriate financial payment of participants, (v) appropriate selection, development, and regulation of challenge strains, (vi) the role (s) of HCS in the licensure of new interventions (e.g., vaccines), and (vii) the potential need for specific research ethics principles/guidelines/frameworks and/or review procedures (e.g., special review committees) for this kind of research.

Although many infectious diseases that have been, or could plausibly be, investigated via HCS are primarily endemic in low- and middle-income countries

(LMICs), the vast majority of HCS have been conducted in high-income countries (HICs). This can perhaps be explained in part by inequitably low levels of research capacity/infrastructure in LMICs and in part by a perceived need to avoid recruitment of vulnerable participants (in LMICs). As a result, less than 1% of HCS participants have been enrolled in LMICs since World War II. This is unfortunate, because (e.g., due to population differences regarding naturally acquiredimmunity, co-infections, genetics, microbiome, nutrition, etc.) research on HIC volunteers may not always be generalisable to the populations in LMICs where many important but neglected diseases are endemic.

There have thus been increased calls for more HCS to be conducted in LMICs in order to produce results that will be more relevant to target populations in endemic settings. This may be particularly important for (i) pathogens that are primarily endemic in LMICs such as those causing many vector-borne diseases (e.g., malaria, dengue, schistosomiasis, Zika, etc.) and diarrhoeal diseases (e.g., cholera, *Shigella*, typhoid, etc.) and (ii) pathogens with a wider global distribution that nevertheless cause a disproportionate disease burden in LMICs (e.g., pneumococcus, *Neisseria*, respiratory syncytial virus, etc.). Although some of the ethical challenges related to HCS may be exacerbated in the context of LMIC populations, it is precisely these populations that are in greatest need of new interventions for neglected infectious diseases.

Despite the potential advantages of conducting HCS in endemic countries, HCS in LMICs may, in addition to more general worries about potentially risky research involving vulnerable human subjects, raise particular ethical and/or regulatory challenges regarding (i) informed consent (due to language barriers and/or limited education of potential participants), (ii) concerns about "undue inducement" (e.g., if financial payment is "too high"), (iii) potential transmission of infection from participants to third parties (e.g., because of the presence of vectors for vector-borne diseases, or the absence of adequate sanitation to prevent the spread of diarrhoeal disease), (iv) the capacity of LMIC regulators and/or ethics committees and (v) the public acceptance of HCS research by local communities.

On the other hand, there may be cases where infection during HCS is less risky/harmful to participants in endemic settings than participants in wealthy developed countries—e.g., if the former has innate (genetic) resistance and/or naturally acquired (partial) immunity to the pathogen under study (making resultant illness less severe), whereas, the latter does not. Participation in HCS may sometimes even have *direct benefits* for healthy participants in endemic, developing countries (which is usually not the case for participants in wealthy developed nations) if/when (i) controlled infection leads to *protective immunity* against endemic diseases that otherwise would have put them at risk and/or (ii) HCS involve infection with locally prevalent pathogen participants would have otherwise likely been infected with later, but *controlled* infection (yielding immunity) leads to *less severe illness* than would otherwise be expected (in light of monitoring of, and care provided to, participants) and/or (iii) HCS involve the testing of a vaccine candidate that results in protective immunity against wild-type infection.

This report maps the terrain of ethical and regulatory issues related to HCS with a focus on studies conducted in endemic LMICs; and it includes an analysis of 13 LMIC HCS case studies of research (published 1992-2018) conducted in Thailand, Colombia, Tanzania, Kenya, and Gabon. It is informed by qualitative interviews involving key stakeholders with expertise relevant to LMIC HCS and a review of relevant bioethical and scientific literature. Based on the views of stakeholders and existing literature, this report concludes by highlighting areas of consensus regarding LMIC HCS, as well as controversial and/or unresolved issues in need of further analyse.

Key Points of Consensus

We found that there was a strong consensus in the literature, and among stakeholders interviewed for this project, that (i) LMIC HCS can be ethically acceptable if there is a sound scientific rationale and studies are conducted to adequate scientific and ethical standards, and that (ii) the scientific rationale for LMIC HCS is particularly strong when they are necessary to produce results that are (more) relevant to the eventual target population for interventions under development. It has even been suggested that (iii) there may be an ethical imperative to conduct appropriately designed LMIC HCS in endemic populations in cases where this would accelerate the development of new interventions that would significantly reduce disease burden in these populations.

There was also consensus that (iv) there is a need to build scientific, ethical review, and regulatory capacity for HCS in LMICs to ensure that HCS can be conducted to high standards; (v) burdens (including risks) to participants and third parties should both be minimised; (vi) high quality informed consent should be assured, e.g., by routinely testing participant comprehension prior to enrolment (which has been common practice in recent LMIC HCS); (vii) researchers should engage with local communities regarding, among other things, the public acceptability of HCS; (viii) participants should receive adequate payment, at least including reimbursement for costs and time; and (ix) children should not be recruited for HCS without significant international consultation and local community engagement/acceptance.

Unresolved Issues

Controversial and/or unresolved issues identified by this project include, first, the question of how the scientific aim to produce generalisable results should be weighed against the protection of participants. Many of the ways in which HCS could be designed to reduce risks to participants might undermine the generalisability of results (e.g., regarding vaccine efficacy). For example, conducting HCS with less virulent strains may be less risky for participants but such studies might not provide reliable information about the wild-type strains of greatest public health importance.

The second set of unresolved issues involves questions about burdens (including risks) and benefits, including (i) the appropriate sharing of the benefits of LMIC HCS with participants and/or local communities; (ii) the acceptable limits to burdens imposed on HCS participants (including the overall upper limit to risk, the conditions under which a low probability of lasting harm is an acceptable risk, and the conditions under which curative treatment may be delayed after an HCS participant is diagnosed with the challenge infection); (iii) the appropriate management of potential risks of participants withdrawing from participation (or absconding) after being infected; (iv) whether the degree of acceptable risk to third parties is different in endemic settings given high local background risk (with some stakeholders viewing third-party risk in such settings as arguably negligible and others suggesting that third-party risks should be minimised whether or not the disease in question is locally endemic). Ultimately, there may be difficult ethical trade-offs between the potential benefits and burdens of different study designs (e.g. while close inpatient monitoring during HCS may reduce the risks of severe infection as compared with outpatient HCS or field trials, inpatient HCS designs may increase other burdens related to the isolation of participants). Systematic risk-benefit assessments that include both scientific and ethical evaluation of study designs may help to optimise the design of HCS and/or vaccine development programs.

The third set of controversial and/or unresolved issues involves questions about selection and payment of participants, for example, (i) the conditions under which HCS should recruit students and/or highly educated individuals (on the one hand, it has been argued that educated individuals are better able to give informed consent; on the other hand, students may feel pressure to participate which could undermine consent and, furthermore, excluding less well-educated individuals may be unfair and/or reduce the generalisability of HCS results to key populations); (ii) the conditions under which studies should recruit individuals with acquired immunity and/or innate resistance to the challenge infection (which might make participation less risky but also affect the generalisability of results); (iii) the conditions under which, if any, HCS involving children would be accepted (by communities and other HCS stakeholders); (iv) whether or not, or the extent to which, it would be appropriate to pay participants beyond reimbursement for costs and time spent participating in a given study.

Fourth, this project identified several unresolved issues regarding the governance of HCS, including (i) whether specific ethical guidelines are needed for HCS and, if so, what their content should be; (ii) whether or not, or the conditions under which, HCS should be subject to special review procedures; and (iii) the appropriate regulation of challenge strains (e.g., whether challenge strains should be subject to existing "Investigational New Drug" regulations, and/or whether regulations should be standardised at the international level).

Conclusion

There are ethical and scientific reasons in favour of conducting LMIC HCS in order to address the persistently high burden of infectious diseases in disadvantaged populations. Careful attention to the ethically salient aspects of HCS design, relevant governance mechanisms, and the acceptability of HCS among participants and communities will help to ensure progress and sustainability of this important area of research which will hopefully produce significant global health benefits. Given the complexities of such studies, and the controversial and/or unresolved issues highlighted in this report, further work is needed by biological scientists, social scientists, and bioethicists to support on-going improvements in the design, conduct, and review of LMIC HCS.

Contents

1   **Introduction**   1
   1.1   Focus of This Report   4
   1.2   Methods...   4
       1.2.1   Literature Review   4
       1.2.2   Qualitative Interviews   4
       1.2.3   Synthesis and Validity Checking   5
   References   6

2   **History of Human Challenge Studies**   9
   2.1   Experimental Infection in the 18th–19th Century   9
   2.2   Early Challenge Studies with Vector-Borne Diseases   11
   2.3   Malariotherapy   14
   2.4   Infamous 20th Century Cases and the Rise of Modern
        Research Ethics   15
   2.5   Late 20th Century...   17
   2.6   Capacity Building in Low- and Middle-Income Countries   19
   References   20

3   **Ethical Issues**   25
   3.1   Intentional Infection   25
   3.2   Benefits   27
       3.2.1   Scientific Rationale and Social Value   27
       3.2.2   Benefit Sharing   37
       3.2.3   Capacity Building   38
       3.2.4   Potential Individual Benefits of Participation
           in Endemic Settings.............................   39
   3.3   Burdens for Participants   41
       3.3.1   Limits to Risk   41
       3.3.2   Minimising Risks   44

	3.3.3	Risks to Participants in Endemic Settings	47
	3.3.4	Long-Term Risks and Lasting Harms	48
	3.3.5	Uncertainty	51
	3.3.6	Other Burdens for Participants	53
	3.3.7	Participant Behaviour	55
3.4	Risks to Third Parties		57
	3.4.1	Third-Party Risks and Studies of Transmissibility	61
3.5	Participant Selection		62
	3.5.1	Vulnerable Populations in Human Challenge Studies	63
	3.5.2	Consent	64
	3.5.3	Education Level	65
	3.5.4	Children	67
3.6	Payment of Participants		71
	3.6.1	Undue Inducement	73
	3.6.2	Other Ethical Issues Related to Payment	75
References			77

4 Community Engagement, Ethics Review, and Regulation 83
4.1	Community Engagement		83
4.2	Ethical Review		85
	4.2.1	Ethical Frameworks for Human Challenge Studies	86
	4.2.2	Potential Models for Special Ethical Review	88
4.3	Regulation		89
	4.3.1	International Regulations	90
	4.3.2	Regulating Challenge Strains	90
	4.3.3	Challenge Studies and Licensure of New Interventions	96
	4.3.4	Regulation of Over-Volunteering	98
	4.3.5	Laws Criminalising Intentional Infection	98
References			99

5 Case Studies: Challenge Studies in Low- and Middle-Income Countries 103
5.1	Cholera and *Shigella* Challenge Studies in Thailand		104
	5.1.1	Rationale and Review Process	104
	5.1.2	Recruitment, Participant Selection, Consent, and Payment	107
	5.1.3	Burdens (Including Risks to Participants and Third Parties)	107
	5.1.4	Summary and Outcomes	108
5.2	Falciparum Malaria Challenge Studies in Africa		108
	5.2.1	Rationale and Review Process	109
	5.2.2	Recruitment, Participant Selection, Consent, and Payment	112

5.2.3   Burdens (Including Risks to Participants
                and Third Parties) 114
        5.2.4   Summary and Outcomes 115
5.3   Vivax Malaria Challenge Studies in Colombia 116
        5.3.1   Rationale and Review Process 116
        5.3.2   Recruitment, Participant Selection, Consent,
                and Payment 117
        5.3.3   Burdens (Including Risks to Participants
                and Third Parties) 120
        5.3.4   Summary and Outcomes 121
5.4   Summary of Case Studies 122
        5.4.1   Rationale and Review Process 122
        5.4.2   Recruitment, Participant Selection, Consent,
                and Payment 123
        5.4.3   Burdens (Including Risks to Participants
                and Third Parties) 123
References 124

6   Conclusions 129
6.1   Lessons Learned to Date................................. 129
6.2   Points of Consensus 130
6.3   Controversies and Unresolved Issues 131
        6.3.1   Burdens and Benefits 132
        6.3.2   Participant Selection and Payment 132
        6.3.3   Governance 133
6.4   Future Directions 133
References 134

Chapter 1
Introduction

Infectious diseases cause a large burden of morbidity and mortality worldwide, with a disproportionately high burden in low- and middle-income countries (LMICs). For example, many vector-borne diseases (e.g., malaria, dengue, schistosomiasis, Zika, etc.) and diarrhoeal diseases (e.g., cholera, *Shigella*, typhoid, etc.) are primarily endemic in LMICs among populations where social, economic, and physiological vulnerabilities are common and/or severe. Furthermore, many other pathogens with a wider global distribution nevertheless cause a disproportionate disease burden in LMICs (e.g., pneumococcus, streptococcus, *Neisseria*, respiratory syncytial virus, etc.). In addition to improvements in public health measures and social determinants of health, new medical interventions (e.g., vaccines and treatments) are urgently needed for many such pathogens. This is because (i) in some cases, no effective interventions exist and (ii) in others, existing interventions are becoming less effective (e.g., due to antimicrobial resistance) and/or are associated with unacceptable risks (e.g., drug toxicities).

Human infection challenge studies (HCS) involve the intentional infection of research participants with pathogens (or other micro-organisms) and primarily aim to (i) test (novel) vaccines and therapeutics, (ii) generate knowledge regarding the natural history of infectious diseases (and/or asymptomatic infection), or (iii) develop "models of infection"—i.e., reliable methods (to be used in studies with aims (i) and/or (ii)) of infecting human research participants with particular pathogens. Modern human infection challenge studies are sometimes referred to as "controlled human infection studies," because they involve *controlling* the selection and/or production of the micro-organism strain(s) and the timing, route, and/or dose of infection; infection in a *controlled* environment; and/or (with the aim to avoid serious harm to research participants) infection with micro-organisms causing no disease or disease that is self-limiting or can be (and is) *controlled* with effective cures or treatments; and/or *controlling* who is being infected (and/or subjected to other experimental interventions) (Selgelid and Jamrozik 2018).

HCS involving reliable models of infection provide an especially powerful scientific method for the testing of vaccines and therapeutics; and they can be

E. Jamrozik and M. J. Selgelid, *Human Challenge Studies in Endemic Settings*, SpringerBriefs in Ethics, https://doi.org/10.1007/978-3-030-41480-1_1

1

substantially smaller, shorter, and less expensive than other kinds of studies. Among other benefits, they can significantly reduce the number of participants that must be exposed to an experimental intervention in order to determine its efficacy. This is because (at least in cases where correlates of protection are unknown) determination of experimental vaccine efficacy requires that a sufficient number of research subjects who receive it, and those (in a comparator arm of a trial) who do not, are actually exposed to—i.e., "challenged" by—the pathogen in question (which may only be a small proportion of participants in field trials).

HCS are commonly used in early stage research for the selection of candidate interventions worthy of further investigation in larger studies. Well-designed HCS can thus lead to significant public health benefits being achieved sooner than would otherwise be possible, meaning that there is sometimes a strong ethical rationale for conducting such studies. In some cases, there may also be strong scientific and ethical justification to conduct HCS in LMIC populations in particular—especially if this would help to generate findings (e.g., regarding immune mechanisms and/or vaccine efficacy) that are more relevant to populations with the highest burden of relevant diseases.

Though numerous infamous historical cases of unethical research involved the intentional infection of human subjects with pathogens, the (sparse) existing bioethical ethical discourse on modern HCS (Miller and Grady 2001; Hope and McMillan 2004; UK Academy of Medical Sciences 2005; Lederer 2008; Miller and Rosenstein 2008; Gutmann and Wagner 2012; Bambery et al. 2015; Shah et al. 2017) appears to reflect consensus that intentional infection of human research participants per se is not ethically unacceptable—whereas grossly unethical challenge studies of the past were wrong for other reasons (e.g., they involved *uncontrolled* infection with especially dangerous and/or deadly pathogens; lack of voluntary informed consent, and sometimes violent force; exceptionally vulnerable populations, such as prisoners; etc.).

HCS are nonetheless ethically sensitive—and, *inter alia*, they raise complex questions concerning (i) the limit of acceptable risks to which healthy volunteers may be exposed, (ii) appropriate financial payment/compensation of participants, (iii) the potential need for special review procedures (e.g., involving dedicated committees and/or the involvement of infectious disease experts), (iv) the need for protection of third-parties from infection (by participants), and (v) appropriate criteria and processes for participant selection/exclusion.

Researchers involved in modern HCS have been especially careful to avoid (severe and/or irreversible) harm to participants, in part via exclusion of vulnerable individuals. This is a major reason why modern HCS have been conducted almost entirely in wealthy developed nations, even for infections/diseases that are usually only present elsewhere. This is unfortunate because—due to population differences regarding naturally acquired immunity, co-infections, genetics, microbiome, nutrition, and so on—research conducted in high-income settings may not always translate well to LMICs where neglected diseases (for which research and development of new interventions are especially important) are endemic. For this and other reasons, there have been increased calls for HCS in endemic settings (Gordon et al. 2017; Baay et al. 2018; Elliott et al. 2018).

In a review of the literature for this project, we identified 13 LMIC HCS published between 1992 and 2018 that were conducted in Thailand, Colombia, Tanzania, Kenya, and Gabon—countries in which the pathogens used (malaria, cholera, and *Shigella*) are endemic. These studies recruited around 400 individuals in total, which is less than 1% of the >40,000 volunteers who have participated in HCS in high income countries (HICs) in since World War II (Kalil et al. 2012; Evers et al. 2015).

Despite the potential scientific benefits of conducting HCS in endemic countries, HCS in such countries may raise (or be perceived to raise) particular challenges regarding informed consent (due to language barriers and/or limited educational background of potential participants) and/or concerns about "undue inducement" (e.g., if financial compensation is "too high", in light of socio-economic status of potential participants) in addition to more general worries about potentially risky research involving vulnerable human subjects. Furthermore, the risks of spreading infections from study participants to third parties may be higher in some endemic areas and/or underprivileged populations (e.g., because of the presence of vectors for vector-borne diseases, or the absence of adequate sanitation to prevent the spread of diarrhoeal disease).

On the other hand, there may be cases where infection during HCS is less risky/harmful to participants in endemic settings than participants in wealthy developed countries—e.g., if the former have naturally acquired (partial) immunity to the pathogen under study (making resultant illness less severe) whereas the latter do not. Participation in HCS may sometimes even have *direct benefits* for healthy participants in endemic countries (which is usually not the case for participants from non-endemic wealthy nations) if/when (i) controlled infection leads to *protective immunity* against endemic diseases that otherwise would have put them at risk and/or (ii) HCS involves infection with a locally prevalent pathogen participants would have otherwise likely been infected with later, but *controlled* infection (yielding immunity) leads to *less severe illness* than would otherwise be expected (in light of monitoring of, and care provided to, participants) (Selgelid and Jamrozik 2018) and/or (iii) HCS participants receive an experimental vaccine that turns out to protect against subsequent infection (after the study) with locally prevalent strains of the pathogen in question.

Weighing the potential benefits and burdens (including risks to participants and third parties) associated with HCS requires careful attention to scientific and ethical aspects of study design, and should also involve learning from local communities regarding local priorities and the public acceptance of potential research designs etc. In any case, HCS should be conducted according to appropriate requirements of ethical and regulatory review, which may need to be adapted to this relatively complex and (in some settings) novel area of scientific research.

1.1   Focus of This Report

This report aims to fill a gap in the current literature by focusing particularly on ethical and regulatory issues that are specific and/or highly salient to challenge studies conducted in LMICs (where many pathogens of interest are primarily endemic). Having reviewed relevant scientific and bioethical literature, constructed case studies of LMIC HCS, and conducted qualitative interviews with relevant experts in LMIC HCS, we ultimately sought to identify (i) areas of consensus regarding ethical issues and regulation in the context of LMIC HCS, as well as (ii) unresolved issues that require further study/analysis.

1.2   Methods

1.2.1   Literature Review

Our review of academic literature and regulatory documents was particularly focused on identifying (i) primary scientific papers detailing LMIC HCS, (ii) relevant historical examples of (other) HCS, (iii) regulatory documents or policy consultations specific to HCS (whether HIC or LMIC), and (iv) bioethical analyses of HCS and/or ethical issues relevant to HCS in LMICs.

Relevant articles published between 1700 and 31st December 2018 were identified through searches in the authors' personal files, Google Scholar, and PubMed. Articles arising from these searches and citations within those articles were reviewed. For LMIC HCS, we included primary publications that gave details of HCS methods and results; conference abstracts were excluded due to lack of detail. Searches were conducted in English and articles published in English were the primary resources. Where articles in other languages had translations of their abstract or article available in English, these were also reviewed. The search strategy included the terms: bioethics, dengue, ethic*, cholera, challenge model, challenge study, controlled human infection model (CHIM), controlled human malaria infection (CHMI), histor*, human challenge, human infection study, malaria, regulat*, schistosomiasis, shigella, typhoid, Zika.

1.2.2   Qualitative Interviews

Our research team conducted qualitative interviews with 45 participants. We initially recruited informants based on involvement in the conduct of recent HCS in LMICs, expertise related to HCS, expertise in research ethics, and/or involvement in the regulation and/or funding of HCS research. Many interviewees currently working in HICs had been involved in and/or had expertise related to LMIC HCS in particular.

Table 1.1 Characteristics of 45 qualitative interview participants

	n	%
Primary area of expertise		
Science	33	73.3
Ethics	7	15.6
Regulatory representative	4	8.9
Funder representative	1	2.2
Primary location of work		
HIC	26	57.8
LMIC	19	42.2
Africa	6	13.3
Asia	9	20.0
North America	15	33.3
South America	4	8.9
UK/Europe	11	24.4
Sex		
Female	20	44.4
Male	25	55.6
Total	45	100

Further informants were recruited via "snowball" sampling, based on suggestions from the above informants at time of interview. As detailed in (Table 1.1), we recruited a diverse group of participants with a wide range of expertise. Deidentified interview transcripts were coded thematically with a combination of pre-set and open coding. The research team, informed by the main aims of the study, agreed upon an initial code list. Coding then progressed openly and iteratively as emergent codes arose and coding categories were further refined as agreed by the research team. Data were organised and cleaned for use in the final analysis. Coded data were analysed to identify overarching themes and sub-themes (that were validated through initial member checking in subsequent interviews and via the mechanisms discussed below) with validated themes being used to inform the structure of this Final Report. As part of the consent processes, interview participants consented to be quoted anonymously (by pseudonym) in this report and other relevant publications and/or to waive the right to anonymity and be quoted by name.

1.2.3   Synthesis and Validity Checking

The findings of the literature review and thematic analyses of qualitative data were synthesised in this Final Report. Draft copies of the Final Report were shared with (i) a subset of participants who provided feedback to the research team (enabling an assessment of internal validity) and (ii) participants at two international meetings of researchers and policymakers with relevant expertise (enabling an assessment of external validity and transferability).[1] Comments were incorporated, in most cases with de-identified acknowledgement in light of participants' wishes.

References

Baay, M.F.D., T.L. Richie, P. Neels, M. Cavaleri, R. Chilengi, D. Diemert, S.L. Hoffman, R. Johnson, B.D. Kirkpatrick, and I. Knezevic. 2018. Human challenge trials in vaccine development, Rockville, MD, USA, September 28–30, 2017. *Biologicals.*

Bambery, B., M. Selgelid, C. Weijer, J. Savulescu, and A.J. Pollard. 2015. Ethical criteria for human challenge studies in infectious diseases. *Public Health Ethics* 9 (1): 92–103.

Elliott, A.M., M. Roestenberg, A. Wajja, C. Opio, F. Angumya, M. Adriko, M. Egesa, S. Gitome, J. Mfutso-Bengo, and P. Bejon. 2018. Ethical and scientific considerations on the establishment of a controlled human infection model for schistosomiasis in Uganda: Report of a stakeholders' meeting held in Entebbe, Uganda. *AAS Open Research* 1.

Evers, D.L., C.B. Fowler, J.T. Mason, and R.K. Mimnall. 2015. Deliberate microbial infection research reveals limitations to current safety protections of healthy human subjects. *Science and Engineering Ethics* 21 (4): 1049–1064.

Gordon, S.B., J. Rylance, A. Luck, K. Jambo, D.M. Ferreira, L. Manda-Taylor, P. Bejon, B. Ngwira, K. Littler, and Z. Seager. 2017. A framework for controlled human infection model (CHIM) studies in Malawi: Report of a Wellcome Trust workshop on CHIM in low income countries held in Blantyre, Malawi. *Wellcome Open Research* 2.

Gutmann, A., and J. Wagner. 2012. Ethically impossible STD research in Guatemala from 1946 to 1948. *Presidential Commission for the Study of Bioethical Issues.*

Hope, T., and J. McMillan. 2004. Challenge studies of human volunteers: Ethical issues. *Journal of Medical Ethics* 30 (1): 110–116.

Kalil, J.A., S.A. Halperin, and J.M. Langley. 2012. Human challenge studies: A review of adequacy of reporting methods and results. *Future Microbiology* 7 (4): 481–495.

Lederer, S.E. 2008. Walter Reed and the yellow fever experiments. In *The Oxford textbook of clinical research ethics*, 9–17.

Miller, F.G., and C. Grady. 2001. The ethical challenge of infection-inducing challenge experiments. *Clinical Infectious Diseases* 33 (7): 1028–1033.

Miller, F.G., and D.L. Rosenstein. 2008. Challenge experiments. In *The Oxford textbook of clinical research ethics*, 273–279.

Selgelid, M.J., and E. Jamrozik. 2018. Ethical challenges posed by human infection challenge studies in endemic settings. *Indian Journal of Medical Ethics.*

Shah, S.K., J. Kimmelman, A.D. Lyerly, H.F. Lynch, F. McCutchan, F.G. Miller, R. Palacios, C. Pardo-Villamizar, and C. Zorilla. 2017. Ethical considerations for Zika virus human challenge trials. *National Institute of Allergy and Infectious Diseases.*

UK Academy of Medical Sciences. 2005. *Microbial challenge studies of human volunteers.* London: Academy of Medical Sciences.

---

[1] We are particularly grateful to participants at the June 2019 (i) Workshop: An ethical framework for human challenge studies (organised by A/Prof. Seema Shah) and (ii) Guidance Development Meeting regarding the ethics of human challenge studies (convened by the WHO Global Health Ethics Unit).

Chapter 2
History of Human Challenge Studies

2.1 Experimental Infection in the 18th–19th Century

The intentional infection of human beings with pathogens with the aim of achieving benefits (chiefly, the prevention of more severe disease) has occurred for centuries; the (semi-)systematic testing and recording of such methods dates to the 18th Century in England (Halsband 1953; Weiss and Esparza 2015). Although the credit for initiating a modern science of vaccination is usually accorded to Edward Jenner (1749–1823), who pioneered the use of cowpox (cow giving rise to the *vache* in vaccine, a term coined by Jenner) to prevent smallpox, variolation (sometimes referred to as 'inoculation', i.e., the prevention of smallpox by injection or insufflation of material believed to produce a mild infection and thus convey an attenuated risk of the disease) began much earlier, in Asia and the Eastern Mediterranean, and was introduced to England and North America in the early 18th Century (Timonius and Woodward 1714; Halsband 1953, Gross and Sepkowitz 1998; Weiss and Esparza 2015). Furthermore, that prior infection with cowpox protected humans against infection with smallpox was widely believed in cattle farming communities in England (and elsewhere) long before Jenner's experiments; at least one English farmer, Benjamin Jesty, is known to have intentionally infected members of his family with cowpox as a means of preventing smallpox in 1774 (i.e., 25 years before Jenner's experiments) (Gross and Sepkowitz 1998). Yet, although there were some earlier 'trials' of smallpox variolation and cowpox vaccination, Jenner's testing of the cowpox vaccine was more systematic, and involved intentional exposure to smallpox after vaccination (with cowpox) to test efficacy in 1796. However, unlike modern human challenge studies, Jenner's investigations in the late 18th Century did not involve (i) systematic study of the methods required to induce disease safely and reliably in humans or (ii) the testing of preventive/therapeutic interventions against a reliable model of infection.

© The Author(s) 2021
E. Jamrozik and M. J. Selgelid, *Human Challenge Studies in Endemic Settings*,
SpringerBriefs in Ethics, https://doi.org/10.1007/978-3-030-41480-1_2

One of Jenner's teachers had been the prominent Scottish surgeon John Hunter, a local pioneer of smallpox variolation (which carried higher risks than the later practice of vaccination) (Turk and Allen 1990). Although Hunter is credited with many positive achievements, he has also become infamous for his attempt to prove his (later falsified) theory that gonorrhoea and syphilis were in fact the same disease. In 1767, Hunter used an experimental (challenge) technique: the injection of "venereal matter" from a patient with gonorrhoea into the penis of a single research subject (Dempster 1978). Though it is sometimes claimed, even in recent times (Gladstein 2005), that Hunter himself was the subject, there is no contemporaneous evidence to support this theory, and it appears more likely that Hunter experimented on another individual—especially since it was known that he had attempted to transmit gonorrhoea to others via inoculation of the skin (Dempster 1978). Importantly, the research subject developed evidence of syphilis, which Hunter took (erroneously) as evidence in favour of his theory (that gonorrhoea and syphilis were the same disease); it now appears more likely that the patient with gonorrhoea from whom the sample was obtained was also infected with syphilis. Thus, the experiment was scientifically flawed and (although mercury-based treatment was provided for the experimental syphilis infection (Wright 1981)) arguably carried significant risks that many would consider unacceptably high—especially on the assumption that this was not a case of self-experimentation (self-experimentation is discussed further below).

During the 19th Century there were significant developments in microbiological understanding of infectious disease. Towards the end of the century in particular, challenge experiments generally became more systematic (discussed below in 'Early challenge studies with vector-borne diseases'). However, from 1800–1880, several experiments that would now be judged highly unethical took place in Europe (Macneill 2010). Irish, German, and Russian physician-investigators injected infectious material from patients with gonorrhoea and syphilis into children and babies, and at least one baby died as a result. Although these investigations did appear to confirm the transmission of such infections, the studies were poorly controlled (due to the rudimentary knowledge of microbiology and lack of available treatments at the time, and perhaps also to the callousness of the investigators). The use of, and harm to, minors (even though some were teenagers who were said to have agreed to participate), furthermore, struck some physicians of the time as immoral and likely unnecessary (Macneill 2010).

One reason it was not necessary to experiment on others (including minors) is that challenge studies can involve self-experimentation, which—especially under such uncontrolled and uncertain conditions—might be considered ethically preferable to the recruitment of others. For example, two scientists deliberately infected themselves with cholera bacteria in 1892. One developed clinical cholera, and this was taken as significant evidence linking the microbe with the disease (Benyajati 1966). Another early challenge study, testing a typhoid vaccine in two 'Officers of the Indian Medical Service' took place in 1896. Though few details are supplied, if these individuals were medically trained and aware of the details of the study then they may have been able to provide (what would now be considered) proper informed consent to the risks to which they were exposed (Wright 1896).

2.2 Early Challenge Studies with Vector-Borne Diseases

In the late 19th and early 20th Century additional early challenge experiments began to occur on a larger scale and (generally) with increasing scientific rigor. Challenge studies investigating what would now be referred to as vector-borne diseases (e.g., yellow fever, malaria, and dengue) were particularly prominent at the time, and often conducted in endemic countries (in contrast to later challenge studies which have been predominantly conducted in non-endemic countries). Famous early examples of such experiments include (i) the failed attempts by Carlos Finlay to transmit yellow fever from symptomatic patients to healthy individuals in Cuba from 1881–1893[1] (Finlay 1886, 1937; Clements and Harbach 2017) and (ii) the successful transmission of malaria via infected mosquitoes in Italy in 1898 by Battista Grassi (Grassi et al. 1898; Capanna 2006). The latter study provided the first experimental evidence that malaria was transmitted to humans by mosquitoes.[2] Since malaria was, at that time, endemic to much of Italy (potentially casting doubt on Grassi's findings, because individuals exposed to mosquitoes during the trial could have contracted the disease elsewhere), a similar experiment was repeated in London (with infected mosquitoes transported from Italy) by Patrick Manson in 1900 (Manson 1900). Manson infected two volunteers (thought to include his son), and successfully cured the induced infection by administration of quinine (Cox 2010).

Elsewhere during the same period, other early research on yellow fever employed challenge study techniques, though such efforts were sometimes unsuccessful and/or harmful. In 1897, Giuseppe Sanarelli, an Italian physician in Uruguay, claimed to have isolated a bacterial cause of yellow fever (now known to be caused by a virus) and injected a culture of these bacteria into 5 hospital patients, perhaps without their knowledge or consent, of whom 3 died (Lederer 2008). The famous Canadian physician William Osler condemned these experiments in the following terms:

> To deliberately inject a poison of known high degree of virulency into a human being, unless you obtain that man's sanction, is not ridiculous, it is criminal. (Sternberg 1898)

Beyond Osler's sentiment regarding the importance of obtaining a person's sanction (i.e., consent) to be challenged, many contemporary readers will additionally object to Sanarelli's apparent callous disregard for the safety of his

---

[1] Although Finlay's overall hypothesis was correct, these experiments failed to demonstrate transmission because the interval between biting infected patients and biting healthy individuals (now known as the 'extrinsic incubation period') was too short.

[2] Ronald Ross, the English contemporary of Grassi who was awarded the 1902 Nobel prize for identifying mosquito transmission of malaria, had (in 1897) shown that parasites were transmitted from human malaria patients to mosquitoes but used challenge studies in birds (with avian malaria), rather than human challenge, to show the transmission *from* mosquitoes.

'research subjects'. While modern HCS standardly involve healthy volunteers, Sanarelli's subjects were neither healthy (being hospital patients) nor volunteers. The virulence of the organism is a more complex issue: the 'poison' in questions was, if not *known* to be highly virulent (since this was being 'tested'), then at least *expected* to be highly virulent (although in retrospect Sanarelli was clearly ill-informed about the infectious agent he thought he was administering). It is noteworthy, however, that other researchers at the time were also using potentially highly virulent forms of the (presumed) agent of yellow fever, albeit with greater care, less harm, and ultimately greater scientific success.

In the same year (1897), in Mexico, a Dr. Ruis injected the blood of yellow fever patients into 3 individuals (whether they consented is unknown) without producing symptoms of disease. Ruis' unsuccessful experiments pre-dated the successful and larger scale studies led by Walter Reed (Reed 1902). In 1900 experiments by Reed and other members of the Yellow Fever Commission in Cuba demonstrated the transmission of yellow fever to healthy volunteers (i) by injection of blood from confirmed cases and (ii) by mosquitoes fed on confirmed cases. Reed's research ultimately led to the development of methods to prevent infection by avoiding mosquitoes (Reed et al. 1900; Lederer 2008; Clements and Harbach 2017).

Reed is frequently credited for establishing a prototype of informed consent for research because subjects in these (what would now be described as) yellow fever challenge studies were asked to sign a contract that outlined the expected risks of the research. The contract also entailed payment of $100 (USD)—the equivalent of more than $3000 in 2019—for two months' participation in the research, which was doubled ($200, equivalent to ~$6000) for those who contracted yellow fever (Lederer 2008; Clements and Harbach 2017). The contract also made note of the high background risk of contracting yellow fever in Cuba at the time, and the relative benefits of high quality medical care for those infected in the course of the research (versus being infected in an uncontrolled fashion with less medical oversight) (Lederer 2008). Although no research subjects in these initial experiments died, the majority contracted yellow fever (in some cases with severe symptoms). One member of the research team (Jesse Lazear) died from yellow fever despite the best available care (Lederer 2008). It has been suggested that Reed's relatively scrupulous proto-consent procedures were motivated by his awareness of earlier criticisms of Sanarelli (Lederer 2008). It has nevertheless been argued that Reed's consent form did not sufficiently emphasise the risk of death due to experimental infection with yellow fever (Chaves-Carballo 2013). Subsequent attempts to develop a yellow fever vaccine in Cuba using a challenge study based on the work of Reed's team (and using a similar consent form) led to three deaths among research participants, public outcry, and the termination of such experiments in Cuba (Chaves-Carballo 2013).

Elsewhere, from 1902 onwards, researchers in Lebanon, Syria, the Philippines, and Australia conducted early challenge studies with dengue virus (which was, at the time, endemic in parts of all four countries, though it has since been eliminated in Australia), which were followed by later studies (from the 1930s onwards) in the USA and Japan (Cleland et al. 1918; Cleland and Bradley 1919; Simmons et al. 1930; Larsen et al. 2015). Early dengue challenge studies sometimes recruited participants who were military personnel and/or medical researchers (including cases of self-experimentation), although Australian researchers also recruited patients from a local asylum (few details regarding recruitment procedures were published), perhaps because of a reported "unexpected difficulty of obtaining volunteers, even with a considerable monetary inducement" (amount not specified) (Cleland and Bradley 1919).

Similarly, early challenge studies of leishmaniasis were conducted in endemic regions of North Africa, the Eastern Mediterranean, and India.[3] In 1910 investigators published results demonstrating that the inoculation of the skin of research subjects with parasites presumed to cause cutaneous leishmaniasis caused local eruption of the disease (though few details regarding the participants were published) (Nicolle and Manceux 1910; Row 1912). In 1921, the transmission of cutaneous leishmaniasis by sand flies was demonstrated in a human challenge study involving self-experimentation (Théodoridès 1997). Two decades later, after multiple failed attempts (Killick-Kendrick 2013), researchers in India demonstrated the transmission of visceral leishmaniasis (kala-azar) to 5 out of 5 healthy volunteers by infected sand fly bite (Swaminath et al. 1942). The researchers were particularly alert to potential challenges of conducting such research in endemic areas, including (i) the potential role of prior immunity among participants from endemic regions (as a result of previous exposure) and (ii) the possibility that participants might be bitten by other insects and/or infected with leishmania during the study. As a result, they recruited volunteers from a nearby non-endemic area, transported them to a research facility in an endemic area where the experimental infection took place under (by the standards of the time) strict isolation from contact with other insects, and returned them to a non-endemic area for longer term observation. Volunteers were also 'generously compensated' with payment of 400 rupees per month (at a time when the usual wage for unskilled labour was less than 200 rupees per month (Palekar 1957)) and provided with curative treatment (Killick-Kendrick 2013). Of note, these human experiments were considered controversial at the time, and previous requests to use prisoners as research subjects were denied (although this may have been in part because local authorities would not permit a reduction in prison sentences as an inducement for inmates to participate (Killick-Kendrick 2013); such inducements had been used in early smallpox vaccine research in the 18th century during which six British prisoners were freed as a reward for participation (Halsband 1953)). Senior British Army officials also refused to approve the use of human participants (multiple animal studies, including challenge studies, were also being conducted) although it has been suggested that the practice was unofficially tolerated, in part because of the significant expected scientific value of the research (Killick-Kendrick 2013).

---

[3] We are grateful to Dr. Kate Emary for pointing us in the direction of early leishmania challenge studies.

2.3 Malariotherapy

The 1927 Nobel Prize in Medicine was awarded to the Austrian psychiatrist Julius Wagner-Jauregg for the discovery of malariotherapy (intentional infection with malaria as treatment) for neurosyphilis,[4] which became a routine treatment in many psychiatric hospitals, administered either by mosquito challenge or by direct injection of human blood infected with malaria (Chopra et al. 1941; Snounou and Pérignon 2013). The use of this 'therapeutic' malaria infection was widespread in Europe, North and South America, and India (Chopra et al. 1941)[5] until the 1940s when penicillin was discovered as an effective treatment for syphilis (Frith 2012; Snounou and Pérignon 2013). The methods used to 'prove' that malariotherapy was effective for neurosyphillis appear quite rudimentary in comparison with 21st Century science (e.g., because the many case series published at the time lacked control subjects), and any attempt to undertake a modern, retrospective review of malariotherapy would inevitably be subject to possible biases, making it difficult to draw firm conclusions (Austin et al. 1992). Some patients died after receiving malariotherapy but, again, it is difficult to know how many of these cases were due to malaria infection itself, as opposed to other factors, including complications of neurosyphilis.[6] In any case, (neuro)psychiatric patients undergoing malariotherapy were effectively used as research subjects by malariologists in de facto human challenge studies that improved scientific understanding of malaria with regards to (i) confirmation that malaria was caused by several different species of *Plasmodium* parasites, (ii) the natural history of disease, (iii) acquired immunity, (iv) transmission dynamics, and (v) the dormant liver stage of vivax malaria[7] (Shortt et al. 1948; Snounou and Pérignon 2013). At least one investigation reportedly involved consent from the patient and his spouse for a study including liver biopsies (Shortt et al. 1948). These malaria challenge studies undertaken as part of malariotherapy (with or without what would now be considered valid consent to research participation) are still cited by HCS researchers today, including in some of the endemic-region malaria HCS reviewed in detail below (Shekalaghe et al. 2014; Vallejo et al. 2016). Psychiatric malariotherapy patients

---

[4]The debilitating end-stage of syphilis that was relatively common at the time and had no effective treatment.

[5]Interestingly (in the context of this review of endemic-region HCS research) while most malariotherapy programs in Europe and North America reportedly used *P. vivax*, at least one Indian centre used *P. falciparum* (which usually causes a more severe form of malaria) because it was believed that the local population had significant immunity to *P. vivax* that would attenuate the benefits of malariotherapy (see Chopra et al. 1941).

[6]However, when falciparum malaria (a more severe form) was used by mistake in malaria-naïve patients for malariotherapy instead of vivax malaria (the milder form of malaria usually used), the mortality rate was much higher (Austin et al. 1992). In contrast, see the use of falciparum in India in the footnote above.

[7]First identified in 1948 when a malariotherapy patient reportedly consented to a liver biopsy (see Shortt et al. 1948)—i.e., in a proto-challenge study.

were, furthermore, the source of parasites used in other studies, including the Stateville Penitentiary program discussed below (Alving et al. 1948; Miller 2013).

While it was not necessary to use malariotherapy patients to study malaria (since at least one similar study with liver biopsy was done contemporaneously in a particularly altruistic healthy volunteer (Shortt et al. 1949)), researchers may have reasoned that malariotherapy patients were ideal candidates for such studies in light of expectations (based on what was known at the time) that they would directly benefit from infection. In retrospect, however, it is questionable whether (i) all patients with neurosyphilis were able to understand and consent to such research (even in cases where consent was sought), and (ii) the persistent use of malariotherapy in the era of penicillin (as a treatment for neurosyphilis) could have been ethically justified.

2.4   Infamous 20th Century Cases and the Rise of Modern Research Ethics

The genesis of modern research ethics (including the development of relevant codes, declarations, guidelines, principles, etc.) is frequently traced to responses to egregious cases of unethical research in the 20th century (Hope and McMillan 2004; Meltzer and Childress 2008). Several of these infamous cases involved intentional infection of research subjects. For example, some of the atrocities committed in the wartime research programs of Germany and Japan during World War II involved intentional infection with pathogens including anthrax, chlamydia, cholera, dysentery, glanders, hantavirus, malaria, paratyphoid, plague, tetanus, tuberculosis, typhoid, and typhus (Tsuchiya 2008; Weindling 2008; Bambery et al. 2015). These programs collectively involved thousands of victims, many of whom died as a result. Prisoners were violently forced to 'participate' (with no option to refuse nor effort to seek consent); participation often involved uncontrolled infection with pathogens known to cause severe disease and sometimes involved the torture and murder of those infected (e.g., by vivisection) (Tsuchiya 2008; Weindling 2008). Despite claims that such research aimed to improve measures to protect military personnel from infectious diseases, much of the 'research' and/or the procedures involved therein did not have a sound scientific rationale and thus would not have been able to inform the development of such measures, even if it had been conducted in a less violent manner (Tsuchiya 2008; Weindling 2008).

In the USA, contemporaneous war-related research also included recruitment of prisoners for infection challenge studies. Although these were conducted under more humane conditions, the voluntariness of consent, and the legitimacy of recruiting prisoners for research more generally, has since been called into question (Bonham and Moreno 2008; Miller 2013). Of particular relevance to current malaria challenge research, American military research during (and after) WWII included the Stateville Penitentiary experiments (discussed in more detail below), which involved infection of prisoners with malaria (including, in later studies, resistant strains of malaria) (Arnold et al. 1961; Miller 2013).

Later (in 1946–48), studies of sexually transmitted infections performed by American researchers in Guatemala involved intentional infection of vulnerable groups (e.g., sex workers, prisoners, soldiers, mentally disabled and institutionalised patients) with pathogens (e.g., bacteria causing syphilis, gonorrhoea, and chancroid) without their knowledge or consent, and also involved deliberately withholding treatment for these infections (Frieden and Collins 2010; Gutmann and Wagner 2012). In the United States, the Willowbrook School study of infectious hepatitis (1950s to 1970s) involved the intentional infection of mentally disabled, institutionalised children with viral hepatitis (which was, at that time, endemic at the school with very high rates of background infection in both 'patients' and staff), with the aim of better describing its natural history, and testing preventive and/or therapeutic interventions (Ward et al. 1958; Rothman 1982; Robinson and Unruh 2008).

Although these studies were eventually met with widespread condemnation, there is a consensus in the (limited) academic research ethics literature (discussed below in more detail) that it was not intentional infection per se that made these studies unethical, but rather other issues, particularly those related to (i) lack of or inadequate informed consent, and/or (ii) exploitation and/or brutal treatment of vulnerable populations (Miller and Grady 2001; Hope and McMillan 2004; Miller and Rosenstein 2008; Bambery et al. 2015).

Nevertheless, in high-income countries, studies did continue among populations sometimes described as 'vulnerable', e.g., prisoners (Glew et al. 1974) and military personnel (among whom, similarly, it may be more difficult to assure truly voluntary informed consent) (Bonham and Moreno 2008). In a retrospective analysis of the Stateville penitentiary malaria challenge studies conducted by the US military, Franklin Miller contrasts this research program with the (other) abusive wartime research discussed above, noting that (although imprisoned) subjects were invited (not forced) to volunteer, carefully screened for health conditions, and monitored closely during the studies—meaning that such research practices would be largely in accordance with many (subsequently developed) codes of research ethics (Miller 2013). Miller does note, however, that during the research severe adverse reactions and one death occurred (the latter reportedly due neither to infection challenge nor to the antimalarial drugs being trialled), raising plausible but unverifiable concerns that researchers may have been more willing to expose prisoners to higher risks because they were incarcerated (and/or because of the perceived urgency of army research that could save the lives of deployed soldiers) (Miller 2013).

By the 1970s, a consensus was building (although it has perhaps never become unanimous) among research ethicists in developed countries that research among such 'captive' groups could be ethically problematic, ultimately resulting in more careful review of research involving military personnel (although this has not necessarily resolved the underlying ethical tensions) (Bonham and Moreno 2008; Miller 2013), and strict regulations regarding research in prisons that eventually curtailed the recruitment of prisoners (Mishkin 2000; Rosenbaum and Sepkowitz 2002).[8] Some have noted that adult students as a group may sometimes share characteristics with other 'captive' groups that might lead to concerns about their ability to consent (although perhaps to a lesser degree)—especially where they are financially or professionally dependent on their academic superiors and/or required to enroll as research participants as part of their studies (Bonham and Moreno 2008). Such considerations may be important for more recent challenge studies, which frequently recruit from student populations.

2.5 Late 20th Century

Later, some post-WWII challenge studies, such as those in the UK Common Cold Unit, involved volunteers from the general population. Although they predated modern ethics regulations, these studies reportedly involved a careful explanation of risks, voluntary consent, and isolation to prevent third party transmission; they did, however, involve risks that were not well characterised at the time, such as the potential for transmission of other pathogens (e.g., in bodily secretions used to administer the infection challenge) for which there were no testing methods available (Tyrrell 1992).

Elsewhere, at least one early (post-WWII) malaria challenge study took place took place in East Africa (an endemic-region), investigating the degree to which sickle cell trait (a genetic condition affecting red blood cells) protects against malaria. In the 1954 publication of this study (Allison 1954), the 30 research subjects are described as adult male volunteers from the Luo people, and it is mentioned that risks were controlled by giving infected subjects "a prolonged course of antimalarial chemotherapy". Few other details apart from the infection rate in sickle cell trait versus non-sickle trait participants are noted—e.g., the publication records neither the presence nor severity of symptoms among participants, nor any consent process. In 1956 there was also a case of self-experimentation by a single investigator in Nigeria who infected himself with Zika virus and attempted to transmit the virus from himself to laboratory mice (Bearcroft 1956).

While it is possible (perhaps likely) that other endemic region/low-resource country challenge studies were conducted between World War II and 1992 (the date of the first case study reviewed later in this report), our review found that there was a very sparse literature regarding endemic region challenge research during this period, especially as compared to the significant and relatively numerous studies published from the late 19th Century to World War II. This may in part be due to the significant social changes that occurred in endemic regions (many of which

---

[8]In one countervailing consideration, Rosenbaum and Sepkowitz (2002) cite a case of a group of prisoners at (US) Jackson State Prison filing an (unsuccessful) lawsuit arguing that prisoners should have more freedom to participate in research, though this may have been partly because of a view that participation in research would entail thorough medical examination and care, which can be difficult for prisoners to access under usual circumstances.

were previously controlled by European imperial powers) at the end of the colonial period.

Ethical concerns (and reactions to the egregious cases discussed above) have perhaps contributed to a reluctance to undertake more HCS in LMICs (in addition to any technical difficulties regarding the availability of necessary laboratory infrastructure etc.) because (i) impoverished individuals and communities may be (perceived to be) particularly vulnerable (e.g., to various kinds of harm, exploitation, inducement by monetary payment, etc.), and (ii) valid informed consent may be (perceived to be) more difficult to assure in some populations within LMICs (e.g., because of language barriers, limited educational background, etc.).

In any case, HCS research has been largely concentrated in HICs, even where such research addresses pathogens that are primarily endemic in LMICs. For example, in the latter half of the 20th Century, North American and European researchers developed malaria challenge models, ultimately leading to several parallel research programs (Spring et al. 2014; Friedman-Klabanoff et al. 2019). At the outset, such studies were subject to few regulatory requirements and/or ethical oversight mechanisms. Parasites were obtained from infected human 'donors'; challenges involved multiple 'wild-type' malaria pathogens (rather than, in the case of falciparum malaria, the few well-characterised laboratory strains in widespread use today); and prisoners and/or army personnel featured prominently among early recruitment of participants (see discussion of these groups above) (Friedman-Klabanoff et al. 2019).

Since the 1980s, improvements in scientific techniques as well as rigorous regulatory and ethical oversight have supported the development of multiple HCS research programs, studying a wide range of pathogens predominantly in HICs (Miller 2013; Darton et al. 2015). Studies collectively enrolling tens of thousands of healthy volunteers (Darton et al. 2015; Evers et al. 2015) have been safely conducted with no deaths and very few serious or lasting harms reported among HCS participants (Roestenberg et al. 2012; Darton et al. 2015). Pathogens/diseases studied in such trials have included adenovirus, BCG (bacille Calmette–Guérin—an attenuated form of *M. bovis* used as a tuberculosis vaccine), campylobacter, *Candida albicans*, cholera, coronaviruses, cryptosporidium, *Cyclospora cayetanensis*, cytomegalovirus, dengue, *E. coli*, giardia, hepatitis A & B, hookworm (*Ancyclostoma caninum* and *Necator americanus*), influenza, gonorrhoea, *H. ducreyi*, *H. pylori*, listeria, malaria, norovirus, parainfluenza, parvovirus, pneumococcus, Q fever, respiratory syncytial virus, rhinovirus, rotavirus, scabies, streptococci (non-pneumococcal), *Shigella spp.*, *Strongyloides spp.*, and typhoid. Overall, not only have the vast majority of HCS been conducted in HICs, but most studies have hitherto focused on pathogens that cause disease in HICs (though they may also affect LMICs), rather than those that are primarily endemic to LMICs; for example, rhinovirus (a cause of the common cold) has been the pathogen associated with by far the greatest number of HCS (at least 55 studies enrolling, collectively, >18,000 participants, in both respects more than for any other pathogen) (Kalil et al. 2012; Darton et al. 2015; Evers et al. 2015). HCS have resulted in unique insights into host-pathogen interactions, as well as the accelerated development of beneficial interventions, including for pathogens primarily endemic in LMICs. For example, HCS have played a role in the development of recently approved and/or licensed vaccines against typhoid (Jin et al. 2017), cholera (Tacket et al. 1999), and malaria (Ballou 2009).

2.6 Capacity Building in Low- and Middle-Income Countries

More recently, there have been calls for more HCS in endemic settings (particularly for pathogens that are primarily endemic in LMICs) in order to accelerate vaccine development and test new interventions in the populations at highest risk of relevant diseases (Gibani et al. 2015; Gordon et al. 2017; Baay et al. 2018). Our review identified no HCS conducted in LMICs from 1956 until 1992 when what appears to be the first LMIC HCS in nearly 40 years took place in Thailand (Suntharasamai et al. 1992). Researchers from LMICs have, however, participated in international HCS projects (where the challenge infection takes place in a HIC). For example, Thai researchers furnished mosquitoes infected with malaria parasites for HCS conducted in the USA (Malaria Vaccine Initiative 2016). However, it has been reported that Thai institutions were initially reluctant to conduct vivax malaria HCS in Thailand (partly because Thai authorities were awaiting further evidence of the safety and utility of the challenge model in question) (Malaria Vaccine Initiative 2016).

From 1992 onwards, Thai researchers successfully conducted challenge studies with cholera and *Shigella* in populations of Thai volunteers. Elsewhere, in the last two decades, well-established research centres in Colombia, Tanzania, Kenya, and Gabon have successfully conducted malaria HCS, often in collaboration with HIC HCS researchers (discussed in detail in Chap. 5). Researchers in more LMICs—including Equatorial Guinea, India, Indonesia, Malawi, Mali, Uganda, and Vietnam—are understood to be considering and/or conducting HCS at present.

Later, we discuss the ethical and scientific case for conducting (more) appropriately designed HCS in endemic LMICs in greater detail (Section "Potential Individual Benefits of Participation in Endemic Settings"). However, even if there is an especially strong case for conducting such studies, certain (other) ethical issues related to their design and conduct warrant particularly careful attention. This is because (i) HCS may sometimes involve, or at least be perceived to involve, particularly high levels of risks (for participants and third parties) and other burdens for participants (and such studies must therefore be carefully designed and conducted to ensure that expected benefits outweigh risks and burdens), and (ii) local and/or international community acceptance of HCS being conducted in endemic LMICs may be contingent on such studies being designed and conducted to especially high ethical (and scientific) standards, and (iii) certain ethical considerations, though familiar in research ethics discourse, may have particular (underexplored) implications in the context of endemic LMIC HCS. The evaluation of these latter implications may both improve the design and conduct of LMIC HCS and/or provide novel case studies relevant to ongoing debates in research ethics. The remainder of this report summarises ethical and regulatory issues relevant to such studies, including insights from stakeholders interviewed for the current project, followed by a comparative review of LMIC HCS published in 1992–2018.

References

Allison, A.C. 1954. Protection afforded by sickle-cell trait against subtertian malarial infection. *British Medical Journal* 1 (4857): 290.

Alving, A.S., B. Craige, T.N. Pullman, C.M. Whorton, R. Jones, and L. Eichelberger. 1948. Procedures used at Stateville penitentiary for the testing of potential antimalarial agents. *The Journal of Clinical Investigation* 27 (3): 2–5.

Arnold, J., A.S. Alving, C.B. Clayman, and R.S. Hochwald. 1961. Induced primaquine resistance in vivax malaria. *Transactions of the Royal Society of Tropical Medicine and Hygiene* 55 (4): 345–350.

Austin, S.C., P.D. Stolley, and T. Lasky. 1992. The history of malariotherapy for neurosyphilis: Modern parallels. *JAMA* 268 (4): 516–519.

Baay, M.F.D., T.L. Richie, P. Neels, M. Cavaleri, R. Chilengi, D. Diemert, S.L. Hoffman, R. Johnson, B.D. Kirkpatrick, and I. Knezevic. 2018. Human challenge trials in vaccine development, Rockville, MD, USA, September 28–30, 2017. *Biologicals*.

Ballou, W.R. 2009. The development of the RTS, S malaria vaccine candidate: Challenges and lessons. *Parasite Immunology* 31 (9): 492–500.

Bambery, B., M. Selgelid, C. Weijer, J. Savulescu, and A.J. Pollard. 2015. Ethical criteria for human challenge studies in infectious diseases. *Public Health Ethics* 9 (1): 92–103.

Bearcroft, W.G.C. 1956. Zika virus infection experimentally induced in a human volunteer. *Transactions of the Royal Society of Tropical Medicine and Hygiene* 50 (5).

Benyajati, C. 1966. Experimental cholera in humans. *British Medical Journal* 1 (5480): 140.

Bonham, V.H., and J.D. Moreno. 2008. Research with captive populations: Prisoners, students, and soldiers.

Capanna, E. 2006. Grassi versus Ross: Who solved the riddle of malaria? *International Microbiology* 9 (1): 69–74.

Chaves-Carballo, E. 2013. Clara Maass, yellow fever and human experimentation. *Military Medicine* 178 (5): 557–562.

Chopra, R.N., B. Sen, and J.C. Gupta. 1941. Induced Malaria with heavy malignant tertian infection. *The Indian Medical Gazette* 76 (6): 350.

Cleland, J.B., and B. Bradley. 1919. Further experiments in the etiology of dengue fever. *Epidemiology and Infection* 18 (3): 217–254.

Cleland, J.B., B. Bradley, and W. McDonald. 1918. Dengue fever in Australia. Its history and clinical course, its experimental transmission by Stegomyia fasciata, and the results of inoculation and other experiments. *Epidemiology and Infection* 16 (4): 317–418.

Clements, A.N., and R.E. Harbach. 2017. History of the discovery of the mode of transmission of yellow fever virus. *Journal of Vector Ecology* 42 (2): 208–222.

Cox, F.E.G. 2010. History of the discovery of the malaria parasites and their vectors. *Parasites and Vectors* 3 (1): 5.

Darton, T.C., C.J. Blohmke, V.S. Moorthy, D.M. Altmann, F.G. Hayden, E.A. Clutterbuck, M.M. Levine, A.V.S. Hill, and A.J. Pollard. 2015. Design, recruitment, and microbiological considerations in human challenge studies. *The Lancet Infectious Diseases* 15 (7): 840–851.

Dempster, W. 1978. Towards a new understanding of John Hunter. *The Lancet* 311 (8059): 316–318.

Evers, D.L., C.B. Fowler, J.T. Mason, and R.K. Mimnall. 2015. Deliberate microbial infection research reveals limitations to current safety protections of healthy human subjects. *Science and Engineering Ethics* 21 (4): 1049–1064.

Finlay, C. 1886. Yellow fever: Its transmission by means of the Culex mosquito. *American Journal of Medical Science* 92: 395–409.

Finlay, C. 1937. The mosquito hypothetically considered as an agent in the transmission of yellow fever poison. *The Yale Journal of Biology and Medicine* 9 (6): 589.

Frieden, T.R., and F.S. Collins. 2010. Intentional infection of vulnerable populations in 1946–1948: Another tragic history lesson. *JAMA* 304 (18): 2063–2064.

Friedman-Klabanoff, D.J., M.B. Laurens, A.A. Berry, M.A. Travassos, M. Adams, K.A. Strauss, B. Shrestha, M.M. Levine, R. Edelman, and K.E. Lyke. 2019. The controlled human Malaria infection experience at the University of Maryland. *The American Journal of Tropical Medicine and Hygiene.* tpmd180476.

Frith, J. 2012. Syphilis-its early history and treatment until penicillin, and the debate on its origins. *Journal of Military and Veterans Health* 20 (4): 49.

Gibani, M.M., C. Jin, T.C. Darton, and A.J. Pollard. 2015. Control of invasive Salmonella disease in Africa: Is there a role for human challenge models? *Clinical Infectious Diseases* 61 (Suppl 4): S266–S271.

Gladstein, J. 2005. Hunter's chancre: Did the surgeon give himself syphilis? *Clinical Infectious Diseases* 41 (1): 128.

Glew, R.H., P.E. Briesch, W.A. Krotoski, P.G. Contacos, and F.A. Neva. 1974. Multidrug-resistant strain of Plasmodium falciparum from Eastern Colombia. *Journal of Infectious Diseases* 129 (4): 385–390.

Gordon, S.B., J. Rylance, A. Luck, K. Jambo, D.M. Ferreira, L. Manda-Taylor, P. Bejon, B. Ngwira, K. Littler, and Z. Seager. 2017. A framework for controlled human infection model (CHIM) studies in Malawi: Report of a Wellcome Trust workshop on CHIM in low income countries held in Blantyre, Malawi. *Wellcome Open Research* 2.

Grassi, B., A. Bignami, and G. Bastianelli. 1898. *Ulteriori ricerche sul ciclo dei parassiti malarici umani nel corpo del zanzarone*, Tip. della R. Accademia dei Lincei.

Gross, C.P., and K.A. Sepkowitz. 1998. The myth of the medical breakthrough: Smallpox, vaccination, and Jenner reconsidered. *International Journal of Infectious Diseases* 3 (1): 54–60.

Gutmann, A., and J. Wagner. 2012. Ethically impossible STD research in Guatemala from 1946 to 1948. *Presidential Commission for the Study of Bioethical Issues*.

Halsband, R. 1953. New light on Lady Mary Wortley Montagu's contribution to inoculation. *Journal of the History of Medicine and Allied Sciences* 390–405.

Hope, T., and J. McMillan. 2004. Challenge studies of human volunteers: Ethical issues. *Journal of Medical Ethics* 30 (1): 110–116.

Jin, C., M.M. Gibani, M. Moore, H.B. Juel, E. Jones, J. Meiring, V. Harris, J. Gardner, A. Nebykova, and S.A. Kerridge. 2017. Efficacy and immunogenicity of a Vi-tetanus toxoid conjugate vaccine in the prevention of typhoid fever using a controlled human infection model of Salmonella Typhi: A randomised controlled, phase 2b trial. *The Lancet*.

Kalil, J.A., S.A. Halperin, and J.M. Langley. 2012. Human challenge studies: A review of adequacy of reporting methods and results. *Future Microbiology* 7 (4): 481–495.

Killick-Kendrick, R. 2013. The race to discover the insect vector of kala-azar: A great saga of tropical medicine 1903–1942. *Bulletin de la Société de pathologie exotique* 106 (2): 131–137.

Larsen, C.P., S.S. Whitehead, and A.P. Durbin. 2015. Dengue human infection models to advance dengue vaccine development. *Vaccine* 33 (50): 7075–7082.

Lederer, S.E. 2008. Walter Reed and the yellow fever experiments. In *The Oxford textbook of clinical research ethics*, 9–17.

Macneill, P.U. 2010. Regulating experimentation in research and medical practice. In *A companion to bioethics*, ed. H. Kuhse, and P. Singer. Blackwell Publishing Ltd.

Malaria Vaccine Initiative. 2016. The challenges of malaria vaccine "challenge" trials: Mosquitoes travel in business class to infect American volunteers—but fare better in economy. https://www. malariavaccine.org/news-events/news/challenges-malaria-vaccine-challenge-trials-mosquitoes-travel-business-class. Accessed 14 Feb 2019.

Manson, P. 1900. Experimental proof of the mosquitomalaria theory. *British Medical Journal* 2 (2074): 949.

Meltzer, L.A., and J.F. Childress. 2008. What is fair participant selection? In *The Oxford textbook of clinical research ethics*, ed. Ezekiel J. Emanuel, Christine Grady, Robert A. Crouch, Reidar K. Lie, Franklin G. Miller, and David Wendler, 377–385. Oxford: Oxford University Press.

Miller, F.G. 2013. The Stateville penitentiary malaria experiments: A case study in retrospective ethical assessment. *Perspectives in Biology and Medicine* 56 (4): 548–567.

Miller, F.G., and C. Grady. 2001. The ethical challenge of infection-inducing challenge experiments. *Clinical Infectious Diseases* 33 (7): 1028–1033.

Miller, F.G., and D.L. Rosenstein. 2008. Challenge experiments. In *The Oxford textbook of clinical research ethics*, 273–279.

Mishkin, B. 2000. Law and public policy in human studies research. *Perspectives in Biology and Medicine* 43 (3): 362–372.

Nicolle, C., and L. Manceux. 1910. Rechereches sur le bouton d'Orient: Cultures, reproduction expérimentale, immunisation. *Annales de L'Institut Pasteur* 9: 673.

Palekar, S.A. 1957. Real Wages in India 1939–50. *Economic Weekly* 151–160.

Reed, W. 1902. Recent researches concerning the etiology, propagation, and prevention of yellow fever, by the United States Army Commission. *Epidemiology and Infection* 2 (2): 101–119.

Reed, W., J. Carroll, A. Agramonte, and J.W. Lazear. 1900. The etiology of yellow fever—A preliminary note. *Public Health Papers and Reports* 26: 37.

Robinson, W.M., and B.T. Unruh. 2008. The hepatitis experiments at the Willowbrook State School. In *The Oxford textbook of clinical research ethics*, 80–85.

Roestenberg, M., G.A. O'Hara, C.J.A. Duncan, J.E. Epstein, N.J. Edwards, A. Scholzen, A.J.A.M. Van der Ven, C.C. Hermsen, A.V.S. Hill, and R.W. Sauerwein. 2012. Comparison of clinical and parasitological data from controlled human malaria infection trials. *PLoS ONE* 7 (6): e38434.

Rosenbaum, J.R., and K.A. Sepkowitz. 2002. Infectious disease experimentation involving human volunteers. *Clinical Infectious Diseases* 34 (7): 963–971.

Rothman, D.J. 1982. Were Tuskegee and Willowbrook 'studies in nature'? *Hastings Center Report* 5–7.

Row, R. 1912. The curative value of Leishmania culture "vaccine" in oriental sore. *British Medical Journal* 1 (2671): 540.

Shekalaghe, S., M. Rutaihwa, P.F. Billingsley, M. Chemba, C.A. Daubenberger, E.R. James, M. Mpina, O.A. Juma, T. Schindler, and E. Huber. 2014. Controlled human malaria infection of Tanzanians by intradermal injection of aseptic, purified, cryopreserved Plasmodium falciparum sporozoites. *The American Journal of Tropical Medicine and Hygiene* 91 (3): 471–480.

Shortt, H.E., P.C.C. Garnham, G. Covell, and P.G. Shute. 1948. Pre-erythrocytic stage of human malaria, Plasmodium vivax. *British Medical Journal* 1 (4550): 547.

Shortt, H.E., N.H. Fairley, G. Covell, P.G. Shute, and P.C.C. Garnham. 1949. Pre-erythrocytic stage of Plasmodium falciparum. *British Medical Journal* 2 (4635): 1006.

Simmons, J.S., J. St, H. Joe, and F.H.K. Reynolds. 1930. Transmission of Dengue fever by Aëdes albopìctus, Skuse. *Philippine Journal of Science* 41 (2).

Snounou, G., and J.-L. Pérignon. 2013. Malariotherapy–insanity at the service of malariology. *Advances in Parasitology (Elsevier).* 81: 223–255.

Spring, M., M. Polhemus, and C. Ockenhouse. 2014. Controlled human malaria infection. *The Journal of Infectious Diseases* 209 (Suppl 2): S40–S45.

Sternberg, G.M. 1898. The bacillus icteroides (Sanarelli) and bacillus X (Sternberg). *Transactions of the Association of American Physicians* 13: 71.

Suntharasamai, P., S. Migasena, U. Vongsthongsri, W. Supanaranond, P. Pitisuttitham, L. Supeeranan, A. Chantra, and S. Naksrisook. 1992. Clinical and bacteriological studies of El Tor cholera after ingestion of known inocula in Thai volunteers. *Vaccine* 10 (8): 502–505.

Swaminath, C., H.E. Shortt, and L. Anderson. 1942. Transmission of Indian kala-azar to man by the bites of Phlebotomus argentipes, Ann. and Brun. *Indian Journal of Medical Research* 123 (3): C473.

Tacket, C.O., M.B. Cohen, S.S. Wasserman, G. Losonsky, S. Livio, K. Kotloff, R. Edelman, J.B. Kaper, S.J. Cryz, and R.A. Giannella. 1999. Randomized, double-blind, placebo-controlled, multicentered trial of the efficacy of a single dose of live oral cholera vaccine CVD 103-HgR in preventing cholera following challenge with Vibrio cholerae O1 El tor inaba three months after vaccination. *Infection and Immunity* 67 (12): 6341–6345.

Théodoridès, J. 1997. Note historique sur la découverte de la transmission de la leishmaniose cutanée par les phlébotomes. *Bulletin de la Societe de Pathologie Exotique* 90: 177–178.

Timonius, E., and J. Woodward. 1714. An account, or history, of the procuring the small pox by incision, or inoculation; as it has for some time been practised at Constantinople. *Philosophical Transactions (1683–1775)* 29: 72–82.

Tsuchiya, T. 2008. The imperial Japanese experiments in China. In *The Oxford textbook of clinical research ethics*, 31–45.

Turk, J., and E. Allen. 1990. The influence of John Hunter's inoculation practice on Edward Jenner's discovery of vaccination against smallpox. *Journal of the Royal Society of Medicine* 83 (4): 266–267.

Tyrrell, D.A.J. 1992. A view from the common cold unit. *Antiviral Research* 18 (2): 105–125.

Vallejo, A.F., J. García, A.B. Amado-Garavito, M. Arévalo-Herrera, and S. Herrera. 2016. Plasmodium vivax gametocyte infectivity in sub-microscopic infections. *Malaria Journal* 15 (1): 48.

Ward, R., S. Krugman, J.P. Giles, A.M. Jacobs, and O. Bodansky. 1958. Infectious hepatitis: Studies of its natural history and prevention. *New England Journal of Medicine* 258 (9): 407–416.

Weindling, P.J. 2008. The Nazi medical experiments. In *The Oxford textbook of clinical research ethics*, 18–30.

Weiss, R.A., and J. Esparza. 2015. The prevention and eradication of smallpox: A commentary on Sloane (1755) 'An account of inoculation'. *Philosophical Transactions of the Royal Society B* 370 (1666): 20140378.

Wright, A.E. 1896. On the association of serous hæmorrhages with conditions of defective blood-coagulability. *The Lancet* 148 (3812): 807–809.

Wright, D. 1981. John Hunter and venereal disease. *Annals of the Royal College of Surgeons of England* 63 (3): 198.

Chapter 3
Ethical Issues

3.1 Intentional Infection

For members of the public, and perhaps many scientists and ethicists, who may be surprised to learn that HCS involving intentional infection (still) take place, the first ethical question may be whether intentionally infecting healthy volunteers as part of research is ever acceptable (Lynch 2012; Evers et al. 2015). Intuitions that such research practices should not be permitted may rest on presumptions that intentional infection with pathogens would involve unacceptably high levels of risk (see Sect. 3.3.1) and/or that physicians should be curing diseases rather causing them to occur among research participants (i.e., that intentionally causing disease is a moral wrong over and above the risks involved) (Hope and McMillan 2004). In an early paper on ethical aspects of HCS, Hope and McMillan provide an extensive philosophical analysis (summarised in the following paragraphs) of the question of whether intentional infection of participants is worse, morally speaking, than imposing other kinds of risks on participants (Hope and McMillan 2004).

Acts resulting in harm are often viewed as less morally acceptable where the harm was intended, rather than merely foreseen. To explore the application of the distinction between intended and foreseen harms to HCS, Hope and McMillan compare infecting a research participant (with a potentially harmful disease) with performing a lumbar puncture on a participant (with the aim of testing the person's cerebrospinal fluid, or CSF) which might generally be considered an acceptable part of research involving human participants. Supposing that both acts might lead to similar risks and/or harms (a lumbar puncture can cause a severe headache, for example, as can many infectious diseases), one might think that only in the HCS case is the harm intended—in the lumbar puncture case, the small probability of severe side-effects might be considered foreseen, but not intended.

There are several possible ways of drawing distinctions between intended and foreseen harms. First, one might argue that foreseen harms are those that occur when the ultimate aim of an act is to bring about a beneficial outcome but there is no (other) way of bringing about this beneficial outcome without bringing about the harm in

E. Jamrozik and M. J. Selgelid, *Human Challenge Studies in Endemic Settings*, SpringerBriefs in Ethics, https://doi.org/10.1007/978-3-030-41480-1_3

question. This way of drawing the distinction might assume that in such cases the (only) *intent* is to produce a good outcome, and any harm that occurs is a *foreseeable but unintended* side-effect. For example, in the lumbar puncture case, there is usually no less harmful way to test research participants' CSF other than by performing a lumbar puncture (i.e. puncturing the lining of the spinal cord with a needle, which itself leads to the risk of side-effects). However, Hope and McMillan note that on this way of drawing the distinction, the harms associated with HCS could also be considered 'merely' foreseen:

> [I]f there were less harmful means of bringing about the testing of the putative vaccine they would be adopted: the risk of harm resulting from infecting the research participants is foreseen but not intended. (Hope and McMillan 2004)

Hope and McMillan further argue that the view that the moral acceptability of an action depends on whether any resulting harms are intended versus (merely) foreseen, where the intention/foresight distinction is drawn in this way, ultimately fails because this would lead to counterintuitive judgements in other scenarios. For example, this view would suggest that it would be acceptable to kill one innocent person if this were the only way of saving more than one other person.

On a second way of drawing the distinction between intended and foreseen harms, one might argue that harms are merely foreseeable when an act that causes a harm in the process of causing a beneficial outcome is more closely causally related to producing the good outcome than it is to producing the harm (Hope and McMillan cite Frances Kamm as articulating a similar idea with her "Principle of Permissible Harm" (Kamm 1989)). However, even if it were the case that this distinction identified an ethically salient difference,[1] it is not clear that the causal connections between intentional infection and harm are any closer than those between a lumbar puncture and the harms of a known side-effect of such a procedure (Hope and McMillan 2004).

Ultimately, Hope and McMillan conclude that such distinctions fail to support ethically relevant differences between the harms of challenge infections and other kinds of harms imposed on healthy volunteers in the pursuit of socially valuable research. Thus, the intentional infection of participants involved in (carefully conducted) HCS can, under certain conditions, be ethically acceptable. The authors note that modern HCS researchers typically take great care to minimise risks to participants (e.g. through the choice of challenge strain and close monitoring, early diagnosis, and treatment of participants). If the risks of HCS are thus within acceptable limits (i.e. such risks fall within widely accepted limits to research risks), and arguments regarding the unacceptability of intentional infection (as compared with other types of research risks) ultimately do not succeed, then there is no *prima facie* reason to rule out the ethical acceptability of intentional infection as a research practice.

---

[1] Hope and McMillan note that intuitions regarding the ethical salience of such 'causal intimacy' might actually arise because of the *probability* of certain harms and benefits arising from an act (since close causal relations between an act and particular consequences are often correlated with a higher probability of those consequences occurring), rather than any other meaningful difference in causation.

Moreover, in the limited ethics literature on HCS to date, there appears to be a consensus that intentional infection per se (and the risk associated with it, once appropriately minimised) can be permissible so long as certain ethical conditions are met (Miller and Grady 2001; Hope and McMillan 2004; Pollard et al. 2012; Bambery et al. 2015; Shah et al. 2017; Selgelid and Jamrozik 2018). As one ethicist from North America noted during an interview for the current project, HCS are not the only kind of research that involve risks to participants without the prospect of benefit:

> There are these arguments [that challenge studies are] contrary to physicians' obligations, [that] *primum non nocere* [the ethical principle that doctors should first do no harm] rules it out and so on. I don't buy those arguments. I think that would rule out most of all research. Research puts people at risk, in part, for the benefit of others.

HCS are nonetheless ethically sensitive and raise important questions, some of which are similar to issues in other areas of research although there may be specific implications of such questions for HCS in particular. Such questions (discussed in the following sections) include those regarding (i) the kinds and level of benefits that would justify exposing healthy volunteers to risk, (ii) how the eventual benefits of HCS should be shared with participants and relevant communities, (iii) the acceptable limit of burdens (including risks) to which healthy volunteers may acceptably be exposed, (iv) the need for protection of third-parties from infection (by participants), (v) fair participant selection/exclusion, especially when considering recruitment or exclusion of vulnerable individuals, (vi) appropriate financial payment of participants, (vii) the potential need for special ethical principles/guidelines/frameworks and/or review procedures (e.g. special committees), (viii) appropriate selection, development, and regulation of challenge strains, (ix) the role(s) of HCS in the licensure of new interventions (e.g., vaccines), and (x) the need for community engagement to raise awareness of challenge studies and ensure that such studies are publicly acceptable to the communities in which they take place.

3.2 Benefits

3.2.1 Scientific Rationale and Social Value

There is a general consensus in research ethics that although research participants may be exposed to risk without the prospect of individual benefit, such risks are only justified to the extent that the study is designed according to a valid scientific rationale that is plausibly expected to lead to significant social value (over and above other study designs involving less risk to human volunteers) (Wikler 2017). The social value of HCS research might include improved scientific knowledge and/or public health benefits, with a key example of the latter being the (accelerated) development of new interventions that reduce the harms caused by the infectious disease being studied (Savulescu 1998; Miller and Grady 2001; Hope and McMillan 2004; Pollard et al. 2012; Bambery et al. 2015).

Thus, where HCS involve particularly significant burdens (e.g., risks to participants and third parties), they arguably require a particularly strong scientific rationale. *Inter alia*, proposed HCS should be compared with other study designs to (insofar as possible) ensure that similar benefits cannot be achieved with research involving fewer burdens. Alternative study designs could include (i) animal models[2]—thus the rationale for HCS might be stronger in cases where the pathogen in question only infects humans and/or (ii) studies of natural history or novel treatments in patients diagnosed as having been naturally infected with the pathogen in question (although natural history studies might sometimes involve withholding proven effective treatment, to which patients might not consent, or which might be ethically problematic for other reasons), and/or (iii) field trials (e.g., where large numbers of individuals are given an investigational vaccine or placebo (or an existing vaccine) and followed until a sufficient number of each group are exposed to the infection in question to make accurate estimates of vaccine efficacy).

With respect to testing interventions, field trials are arguably the most important comparator for HCS designs. Compared to field trials, HCS may often be shorter, less costly, and involve fewer participants—thus allowing a more efficient 'selection' of potentially effective interventions which could (if this results in faster development, licensure, and implementation of such interventions) lead to public health benefits being achieved sooner (Sauerwein et al. 2011; Roestenberg et al. 2018b, c) (see also Sect. 4.3.3). Where this is the case, there may thus be a strong ethical rationale to pursue challenge studies as part of vaccine development.

Furthermore, other things being equal, there may be a stronger scientific and/or ethical rationale to conduct challenge studies where (i) field studies are impractical (e.g., where a pathogen is often asymptomatic and/or currently causes few cases but is likely to cause sporadic epidemics in future (Shah et al. 2017)) and/or (ii) immune correlates of protection are unknown (since, if it is already known that a certain measurable immune response to a vaccine is correlated with a certain level of clinical protection, then there would be arguably be less rationale for infecting participants in order to measure vaccine efficacy), and/or (iii) there is a clear pathway from HCS to licensure (discussed later in Sect. 4.3) (Chattopadhyay and Pratt 2017).

Thus far, the advantages of challenge studies in terms of reduced time and reduced costs (etc.) have rarely been quantified. However, there are clearly quantifiable advantages in terms of the number of participants—challenge studies typically enroll less than 100 participants (see the Case Studies reviewed in Chap. 5. of this report) or sometimes up to a few hundred, whereas field trials of vaccines typically enroll thousands (or sometimes many tens of thousands) of participants. To take one comparison, the recent HCS that supported WHO pre-approval of a new typhoid vaccine challenged 103 participants with typhoid (Jin et al. 2017), whereas previous field trials of other typhoid vaccines have

---

[2] Animal models, including animal challenge studies (perhaps especially in non-human primates), may raise ethical issues of their own, although these are not the focus of this report.

involved between tens of thousands and up to around 100,000 participants (Wahdan et al. 1980; Levine et al. 1987). On the other hand, although vaccine field trials involve risks for participants, including risks of exposure to the pathogen in question (perhaps especially among groups who receive a placebo vaccine in cases where the investigational vaccine is shown to provide individual protection), they typically involve less burdens per participant than HCS (which, for example, generally require much closer monitoring of participants, especially during inpatient HCS), and do not necessarily involve a high probability of infection (whereas HCS participants are intentionally exposed to infection, usually in such a way as to ensure that most, if not all, become infected). Thus, risk-benefit assessments comparing HCS with other designs will require both scientific and ethical analysis. Ideally, researchers and funders would be able to specify the expected benefits, expected burdens, and expected costs of all plausible potential research programs (e.g., HCS vs. field trial) aimed at particular scientific knowledge and/or public health benefits, so as to permit the rational and efficient design of research programs and the efficient allocation of resources. This can be difficult, especially when the results of different scientific investigations, and the probability and/or benefit of eventual implementation of a new intervention, are difficult to estimate *ex ante* (and often overestimated even by experts (Kiwanuka et al. 2018)) in part because of uncertainties regarding the outcomes of scientific research (see Box 3.1), and in part because of uncertainties about the translation of (early phase) results (including via regulatory development pathways) into the licensure and deployment of novel interventions leading ultimately to public health benefits. Yet recent analyses of the role of HCS in vaccine development have suggested developing systematic algorithms that could help to determine the optimal development strategy and/or series of studies required (Roestenberg et al. 2018). Stakeholders interviewed for this project were optimistic regarding the potential for well-designed HCS to accelerate the development of interventions while involving fewer participants,[3] although some were more cautious regarding efficiency and/or cost-effectiveness comparisons between HCS and field trials (see Box 3.1).

Box 3.1 Rationale for HCS

[I]f you have a good model … then I think it can be very useful in streamlining vaccine development and … eliminating candidates that [are] not that great, and ensuring [that] those candidates that have the best safety and efficacy profile move forward … [You] could also translate that to therapeutics … if you have a model that can be used for therapeutics, you can down select candidates in a much more efficient and less costly manner. Prof. Anna Durbin, scientist, USA

---

[3]N.b. the interview sample for this report was potentially biased by the preferential inclusion of scientists active in HCS research—see Sect. 1.2.

[HCS] will try to mimic as best as they can what happens when you've got a real infection with the pathogens in order to detect ... the ability of your vaccine to protect, and you could do this with a limited amount of subjects ... involved and not requiring very large field trials in order to get information on the effectiveness of your vaccine. [Regulatory representative]

[T]he essential question that needs to be asked is: If we pursue infection challenge studies, how much marginal value are we getting from that pursuit against the [alternative] of pursuing that same acquisition of knowledge by another means? ... It can be really, really hard to know which scientific method is going to be the most expedient in arriving at a given state of knowledge ... and often we're not as good at making those predictions as you would think ... and I think part of the reason why we're not good at that is [that] there are a lot of different conditions that need to hold in order for a particular research study to yield the kind of social benefits that [the] research study is directed towards. [Jonathan Kimmelman, ethicist, Canada]

[T]he point about cost effectiveness is really interesting ... undoubtedly doing a challenge trial of a vaccine is going to be quicker and cheaper than doing a large phase 3 field trial that runs over several years—but it doesn't mean that [challenge studies are] as quick or as cheap as you might like ...

[T]here are large costs associated with [them] financially and in terms of time, but yeah they can certainly accelerate that process through but that relies on people actually believing the data in the challenge model. [Malick Gibani, scientist, UK]

3.2.1.1 Generalisability

A key link between the scientific rationale and the social value of a study is the generalisability of the findings (whether these consist of knowledge of disease pathogenesis or estimates of the efficacy of a novel intervention) to the population and context for which the eventual benefits of a research program are intended (Wenner 2015, 2017). The ethical acceptability of LMIC HCS might thus be contingent on such studies generating scientific knowledge that is particularly relevant to LMICs (i.e., more relevant than knowledge that could be generated by HCS in HIC study populations) and/or testing interventions that would (if found to be effective) be particularly beneficial to communities in LMICs (Selgelid and Jamrozik 2018, Wenner 2015, 2017).

Certain choices in study design (e.g., regarding the selection of participants, choice of challenge strain, method of challenge, etc.) might lead to results that are more or less generalisable to the populations most at risk of a particular disease. For example, it would be ethically preferable to conduct HCS in LMICs if/when these can provide results that are more generalisable to (LMIC) populations at risk than results generated by HIC HCS. This might often be the case because of features shared by potential (LMIC) study participants and the (LMIC) population(s) most at risk of the infection in question (e.g., in terms of naturally acquired immunity, co-infections, genetics, microbiome, nutrition, etc.). LMIC HCS in such circumstances would, other things being equal, arguably have a stronger scientific and ethical rationale than a similar HCS design conducted in a (non-endemic) HIC population (see Section "Potential Ethical Imperative for Challenge Studies in Endemic Settings"). Yet there may be exceptions to this general pattern; such inferences are complex and require careful assessment—especially, for example, where HCS in adult volunteers are intended to be generalisable to children in endemic settings (see Sect. 3.5.4.1).

The choice of challenge strain is another element of study design that might influence generalisability and thus give rise to ethical tensions. Using an attenuated challenge strain (or a single well-characterised strain whereas natural infection involves multiple strains), for example, might reduce the risks to participants. If using such a strain means that the findings are not generalisable to wild-type infection (i.e., do not accurately predict the efficacy of a vaccine) in the relevant population and/or epidemiological context, however, then this could undermine the ethical rationale for using HCS rather than an alternative study design (Chattopadhyay and Pratt 2017; Selgelid and Jamrozik 2018). Thus, although the risks to participants associated with the use of wild-type strains in HCS (such as those used in the Colombian vivax program reviewed below or those used in the pivotal trial supporting recent WHO approval of a typhoid vaccine (Jin et al. 2017)) require careful justification, the use of such strains may sometimes be more ethically justifiable than the use of attenuated strains (or a strain that does not adequately reflect the complexity of natural infection) (see Boxes 3.2 and 3.3). Many stakeholders interviewed were well aware of these complexities, although it was often acknowledged that more data are needed (e.g., comparing HCS findings with subsequent field trials) to inform judgements regarding generalisability. In at least one case, concerns regarding a lack of generalisability (from the challenge strain to local malaria strains) reportedly led to the abandonment of a proposed HCS in India (see Box 3.3).

Box 3.2 Generalisability of HCS

[A] challenge trial … produces information that's very controlled and so translating that to actually knowing about a vaccine's efficacy … in a population that's previously been exposed to other infections or people who are infected by mosquitoes instead of … intravenously, those things can be difficult … [J]ust knowing what the value is of a challenge trial and how likely it is to lead to licensure is sometimes complicated. [Ethicist, North America]

Challenge studies are a model for real infections. They are not real infections, in the sense that you generally have to manipulate the dose to get higher attack rates, [and] you [often] do a different type of monitoring than would happen in the field and so you are dealing with a very different setting from wild infections. [Scientist, UK/Europe]

The problem is that [HCS are] a bit artificial in the sense that … you cannot really give the right type organism because it will be too virulent and it will be ethically challengeable, so the [end result] is that you end up [attenuating] your [challenge] pathogen [to the point] where it is [not very] virulent [and then] you may wonder what is the value [of the result of the study], what is the real efficacy of your vaccine? The trade-off between these two extremes … is one of the key factors that will tell us whether a human challenge can be helpful in the beginning of the development [of a] vaccine. [Regulatory representative]

If that challenge is not really giving you a good readout on who that end target population is, [then] you're potentially putting people at risk [without a good justification and] getting an answer that's not going help you decide whether it works or not in that target population. [Scientist, North America]

I think what the [dengue HCS] community is saying is that they're developing attenuated strains [for use in HCS] … but then, what is the relevance? How, how can you justify [it] if it's attenuated to the point it's so different to [wild-type] dengue infection? How do you know that it's relevant to protection against dengue? [Scientist, UK/Europe]

Box 3.3 Generalisability of malaria HCS

Each strain of each malaria species is different, so if we show [vaccine-induced] protection against [a particular] HCS strain this might not lead to natural protection [against wild-type strains]. [Scientist, Asia]

[With vivax malaria HCS] you're using natural strains. That's …a problem and it's an advantage. [It is] a problem because you don't really know exactly which parasites you're giving to the volunteers, at least in the first instance, if you're doing mosquito challenge, because you have to [feed] the field mosquitoes on patients [who have] a natural infection. [On the other hand, using such] wild type [parasites from natural infection] is an advantage in terms of how you test your vaccine. The results of your vaccine [trial] are much more real world [because] you're exposing your vaccinated volunteers to the same challenge as if they were walking through a jungle in an endemic area. [Scientist, Asia]

[A] malaria challenge study … was taken to a committee constituted by the Indian Council of Medical Research. That committee told them that, you know, in principle, all of this is fine but to make this study more epidemiologically relevant you must use a challenge strain from India. I don't think they understood what it takes to make a challenge strain. And how difficult it would be for the vaccine developers to come up with the challenge strain. That did not move forward because it was just too complicated. [Gagandeep Kang, scientist, India]

3.2.1.2 Potential Ethical Imperative for Challenge Studies

Given that HCS are often expected to lead to significant public health benefits more efficiently than alternative research designs, some commentators have argued that there is an ethical imperative to conduct HCS if (or when) no other (less burdensome) feasible research design could obtain similar results and/or if (or when) *not* performing HCS could lead to greater net harms including (i) longer delays to the development and implementation of beneficial new interventions for (neglected) infectious diseases and/or (ii) the exposure of more participants to potentially greater risks in alternative study designs (e.g., field trials involving larger numbers of participants exposed to risks associated with experimental vaccines, where similar vaccine efficacy estimates could have been more efficiently obtained with fewer participants in a challenge study) (Bambery et al. 2015; Selgelid and Jamrozik 2018). Assuming useful HCS can be conducted safely (and other general ethical requirements are satisfied), they might, in such circumstances, be arguably not only ethically permissible, but ethically required or obligatory (i.e., there would be an ethical imperative to conduct such trials) (Bambery et al. 2015; Selgelid and Jamrozik 2018). While few interviewees endorsed such a claim without qualification, there was widespread agreement that the ethical (and scientific) rationale for certain HCS designs in particular circumstances (including in LMIC populations) could be particularly strong, and that pursuing alternative study designs in certain circumstances might be less ethically acceptable (see Box 3.4).

Box 3.4 Could HCS sometimes be ethically obligatory?

There may be a case where it is impossible to test a vaccine for some disease that we can anticipate will arise and it is within the bounds of reasonable risk to do a human challenge study. Then I think it might be pretty close to something I'd call ethically obligatory, if not using that exact terminology. [Ethicist, North America]

I think where it … almost ethically would be obligatory is if you had a disease where the conventional, old way of doing the testing was going to take five years and then the placebo control group people were going to get sick and die, but you could do a human challenge study [and] it'd be done in a month instead of recruiting people over a long haul. [Scientist, North America]

we actually thought that *Shigella* was an example of a case where… human challenge studies are essentially the only responsible way to proceed with vaccine development. To think that… one could immunise thousands of toddlers and then await disease occurrence and incidence [in a field trial with] an experimental vaccine that hasn't been well characterised, you know, that's also hard to justify when you can do a challenge model in a handful of consenting adults [so that] some other issues can be addressed before you take it out to the target population for its final efficacy studies. I almost think it's obligatory to do challenge studies. [Carl Mason, scientist, USA]

Potential Ethical Imperative for Challenge Studies in Endemic Settings

The vast majority of HCS have been conducted in high-income countries. Over 40,000 people have participated in HCS in the ~70 years since World War II (Evers et al. 2015), yet the 13 LMIC case studies we review below enrolled a total of around 400 participants—i.e., less than 1% of the global total. Even HCS research on pathogens that are primarily endemic in LMICs has been largely conducted in non-endemic HICs. Potential reasons for this include: (i) the presence of more/better funded research infrastructure and researchers in HICs (Baay et al. 2018), (ii) the availability of healthcare resources in HICs that can be devoted to caring for HCS participants, thus providing greater assurance of risk minimisation (Gordon et al. 2017), (iii) the view that recruitment among vulnerable populations should be restricted to types of research that are likely to lead to novel medical interventions that could be shared, ideally in the near future, with the local population—this would thus rule out HCS except where these involve or (in the case of HCS aimed at infection model development) facilitate testing drugs or vaccines (Macklin 2003; Pratt et al. 2012; Wenner 2017), and (iv) regulations and/or norms in LMICs that sometimes require prior testing (e.g., early phase studies) of an intervention or research model in the country of the sponsor of the research (usually in a HIC) before testing in an LMIC (Malaria Vaccine Initiative 2016).

The relative current research capacities of HICs and LMICs, including capacity for HCS, are arguably the result of longstanding injustices in the global distribution of wealth and thus funding for research. This has in turn contributed to a relative neglect of research regarding pathogens that are mainly endemic in LMICs and the perpetuation of large inequities in the global burden of disease. There have thus been calls for more HCS in LMICs (Gibani et al. 2015; Gordon et al. 2017; Baay et al. 2018). Furthermore, If there is an ethical imperative to conduct HCS (in general) that is grounded (in part) in the need to relieve significant burdens of infectious disease, then there is arguably an even stronger ethical imperative to conduct (appropriately designed) HCS in LMICs in particular because, *inter alia*, (i) LMIC infectious disease research (including HCS) could lead to significant public health and/or research capacity-building benefits where they are most needed, (ii) HCS may be more efficient and thus lead to such public health benefits more quickly and/or cost-effectively than other study designs, (iii) HCS performed in HIC populations (and estimates of efficacy of any interventions tested there) may not always be generalisable to LMIC populations, and (iv) studies of acquired immunity (which, it is hoped, will improve vaccine development for neglected pathogens) must recruit previously infected individuals who (for many relevant pathogens) live predominantly in LMICs (Selgelid and Jamrozik 2018). In some cases, the ready availability of pathogen strains from locally infected patients may also be an advantage (e.g., for vivax malaria, which currently cannot be maintained in a laboratory culture), as compared with the logistical complexity and costs of exporting these strains to HICs for HCS and other research (Malaria Vaccine

Initiative 2016). Interviewees for the current project raised multiple reasons why it might be advantageous to conduct HCS in LMICs (see Box 3.5).

Box 3.5 Rationale for HCS in LMICs

[I]t makes sense to bring a challenge model to evaluate ... an intervention in the population where it will ultimately be used because you'll get the best handle of whether it's going to work or not ... and, even if it doesn't work, being able to analyse the reasons why it failed might help you to do better the next time around. [Gagandeep Kang, scientist, India]

[T]he main scientific justification for doing it in endemic regions [is] that you are testing vaccines in the population where they will be deployed. [I]t's just the correct model, which always [involves doing the research in] the correct study population. Because it is actually the population where you'll be using the vaccine. [Scientist, Asia]

I don't think it's in the best interest of low-resource countries [to require that HCS are first conducted in HICs before moving to LMICs] because, if we start with those stipulations, many of the things that are needed by low-resource settings will never be studied. [Ethicist, North America]

If you do a challenge study in an endemic setting, it's viewed as responsive to the health needs of the people in that country in addressing a problem they understand and think is important. There may be a baseline risk of transmission that people already accept or just are aware of, and so introducing an additional risk doesn't change the risk profile for them greatly in their estimation. So I think these are the types of reasons people give as saying it might be better, ethically, to do a study in an endemic setting. [Ethicist, North America]

Not only are we opening up research possibilities for African institutions, we're actually [now] able now to address research questions that were off the table before this happened ... because you can only ask these questions in a malaria-exposed population. [Scientist, North America]

Most of the malaria challenge models have been conducted in non-endemic settings and that's problematic from a scientific point of view, in that you don't necessarily know whether or not the volunteers are representative of the likely final beneficiaries of a vaccine ... either in terms of genetics or in terms of acquired immunity. [Scientist, Asia]

Much malaria vaccine work was also done using human challenge studies in the US and ... you can set up a perfectly competent, capable study group in sub-Saharan Africa that would be able to do it in endemic areas and answer many of the questions that you really need to have answered much more quickly. [Scientist, North America]

I think [challenge studies in endemic settings are] great idea and... I think, in fact, maybe justice demands it. ... the notion of ... distributive justice in research ... says that as early as it's ... reasonably safe to do so, we ought be doing these studies in the population that stands to benefit. I mean ... that's what justifies the burden it seems to me, in part. [Ethicist, North America]

Some interviewees even questioned whether continuing to conduct HCS in HICs for pathogens primarily endemic in LMICs could (still) be scientifically and/or ethically justified (see Box 3.6).[4] On the other hand, some participants noted that both HIC and LMIC HCS could be justified, depending on the scientific question being investigated. For example, as discussed above, some HCS aim to test vaccines intended for use among young children who often have no immunity to the pathogen in question (see Sect. 3.5.4). In such cases, even if the eventual target population for the vaccine were children living in endemic areas of LMICs, adult HCS participants in HICs may be a useful model, at least in respect of their level of pre-existing immunity, since (like young children in LMICs) they have no immunity from prior exposure, whereas many of the adults living in endemic areas will already be (semi-)immune, and thus efficacy trials in the latter group would not necessarily be highly generalisable to children (see Box 3.7).

Box 3.6 On-going justification of HCS in HICs

I don't understand why you wouldn't … do challenge studies in endemic countries? I mean I would turn it around and say – why are we doing them in non-endemic countries? [I think] research should be focused in the settings in which those health problems are occurring and I think this is a colonial history that we have which has been propagated by institutions [in high income countries]. The current situation is that they're the ones who have the facilities to do these. They have more funding to do this kind of work, so they just do it where they are as opposed to where it's needed. [Scientist, UK/Europe]

[Starting HCS research in HICs was the most] pragmatic way but I'm starting to question why we haven't stopped. [C]onducting research in Oxford with all those students, giving them so much money? I wonder if that's a good thing to do. I would question that. So rather than questioning: why here [in] an endemic setting? – Why Bangkok? – I want to question: why Oxford? [Scientist, Asia]

I think it is, it's almost unethical to conduct [HCS] in non-endemic settings because those subjects do not benefit at all. They aren't going to be exposed to the disease. Their immune responses aren't typical. The genetic makeup is different. For a whole variety of factors. If you have a viable vaccine or you think it's a viable vaccine, it should be evaluated in endemic settings and, and brought to market as quickly as possible to benefit those people. And conducting studies, non-endemic studies only delays things. [Scientist, North America]

[F]or years, we've been doing CHIM studies in Maryland and Oxford where as it turns out there isn't a lot of malaria … I get the reasons to do that, that you need facilities with all kinds of sophisticated technology and you need the expertise … [but] morally there's an argument for doing these studies in low resource settings as soon as possible and indeed [the same argument] suggests that perhaps there's something even more problematic about doing them in settings where nobody's … going to be exposed the risk of disease. [Ethicist, North America]

---

[4] Although note again that the interview sample was potentially biased by the preferential inclusion of researchers with an interest in LMIC HCS.

Box 3.7 The need for both LMIC and HIC HCS

[T]here are certainly scientific arguments why one might choose different populations but I don't think, as some people have argued, that, scientifically, it is, by definition, better to be in an endemic setting. I don't think that's true. I think, scientifically, there are good arguments for doing things in developing countries but ... the case isn't always in my view adequately made, for why that is better, or not, [than in] a developed country. [Scientist, UK/Europe]

[Testing interventions for which LMIC children are the target population] is the one area [in which it might be justified to do] the challenge studies in non-endemic countries because ... the most practical way of providing data that is relevant to young children, at least in the case of malaria, is [to conduct studies with] adults [in non-endemic settings], because [like children] they're non-immune. [Scientist, UK/Europe]

[Y]ou could argue that ... testing vaccines in malaria naïve adults in Oxford is more predictive of the vaccine efficacy in ... endemic regions ... but we already know that children growing up in endemic regions are exposed to other parasites [and] those infections modulate their immune responses, there's lots of other factors that impact their immune response so even in that setting you can't predict the efficacy on the ground. [Scientist, UK/Europe]

3.2.2 Benefit Sharing

The topic of benefit sharing has been a prominent focus of international research ethics discourse during recent decades (El Setouhy et al. 2004; Njue et al. 2014; Wenner 2015). The sharing of benefits with study participants and/or local populations is commonly cited as an ethical requirement for international research involving human participants, but controversy surrounds questions regarding the nature, content, and weight of such a requirement (El Setouhy et al. 2004; Wenner 2015, 2017). How such a requirement is understood has important implications for the ethical justification of HCS conducted in LMICs.

For example, a requirement for tangible benefits, such as drugs or vaccines approved as a result of a study, to be made available to the local population can lead to a reluctance to conduct basic science and/or *early phase* research (including some HCS designs) in LMICs because such research is not intended to lead to the immediate development or approval of such interventions (Macklin 2003; Wenner 2015, 2017). Nevertheless, such research can lead to benefits in terms of scientific knowledge that is necessary for the development of such interventions in future. In particular, it can lead to knowledge that is of particular relevance to the population in question (Wenner 2015, 2017). Indeed, many of the LMIC HCS reviewed later in this report were focused on model development (where such models may be used later for the testing of interventions) and/or understanding locally relevant aspects of host-pathogen interaction. It may thus be more relevant to consider the overall long-term expected benefits of a program of research as a whole (rather than a single study in isolation) (London and Kimmelman 2019) so long as there is reasonable confidence that the program can/will continue—and this is unfortunately not always the case, because a lack of funding or other issues may delay or entirely halt a proposed research program (see Box 3.8).

Box 3.8 Benefit sharing

[T]hose subjects who are involved in these studies should be able to benefit from the vaccine that is eventually licensed. So you should be focusing on doing the studies in the target population. [Scientist, North America]

Often I find [that] much of … research ethics is preoccupied with a synchronic view of clinical research looking merely at a single clinical trial as opposed to the relationship of that investigation with subsequent investigations, and/or investigations that preceded it. [Jonathan Kimmelman, ethicist, Canada]

I wish we had been part of a larger development plan … because of our bureaucracy, our trials were slow and [because funding was no longer available] we were unable to really conduct the studies that we needed or probably should have conducted or would have been of important value … [O]nce we had established a good challenge model, it would have been helpful and probably more ethical … to be able to continue working [and to test (more) vaccines]. [Carl Mason, scientist, USA]

On the other hand, if benefit sharing requirements apply to vulnerable populations in HICs, and these populations are being recruited for HCS with pathogens that are not locally endemic, then there might be no benefits that could meaningfully be shared with such populations, which implies that it may be more ethical to conduct such studies among vulnerable populations in LMICs than among those in HICs (see Sect. 3.5.1).

3.2.3 Capacity Building

Research may lead to other kinds of ancillary benefits (not directly related to the social value of answering a research question), including (i) research capacity building, (ii) other assistance to the local community, (iii) ancillary benefits to individual participants (e.g., healthcare, provision of testing, etc.), and (iv) payment of participants (discussed in Sect. 3.6). Ethics committees usually exclude these benefits from the risk-benefit assessment of individual studies but whether that is appropriate is controversial and in any case it has been argued that such benefits at least be considered as part of the justification for research programs in LMICs (El Setouhy et al. 2004).

Research capacity building might include contributions to relevant infrastructure, equipment, training of scientific staff, and training of ethics review and/or regulatory body members. Many of those interviewed as part of this project saw building capacity for HCS (and other research) in the countries in which relevant diseases are endemic as important. In addition to technical/scientific capacity building, participants identified ethics capacity building as a particular area of need because many ethics committee members (in LMICs) were not familiar with HCS designs and required significant training and engagement (see also Sect. 4.1) to facilitate review of recent LMIC HCS (see Box 3.9).

Box 3.9 Capacity building

[W]e've been receiving vaccines and drugs developed from other places, so some other people have gone through this to enable us to have what we have now. Now that we are building our capacity, it is also our time now to possibly start … helping and building the next generation of vaccines and drugs. [Scientist, Africa]

[E]ngaging the investigators in developing countries, and having them involved at the beginning, at the ground level of things, and having some of those research resources distributed into places … where the research can be done locally [is] beneficial to the local research community and … can sometimes be a source of pride for the country itself. [Scientist, North America]

I do think capacity-building in endemic settings and making sure that other countries have the ability to do research that they value and that their communities value is really important, and that, … just having collaborations that are global is not sufficient, that we really do need to be thinking more about building capacity. And so it could be that building capacity in human challenge studies will be important to help countries do things … they really value and think are important. [Ethicist, North America]

The issue is how do we capacity build the ethics of your committees to address the new changes that are coming in … proactively, not wait for things to happen, for them to catch up with how they review … I think there's a lot of experience in ethical review, there's a lot of capacity building that has been going on. [Scientist, Africa]

One of the things we did when we are setting up the platform was to [hold] joint meetings between the scientists and the ethics committees, so that people are [able to] share their experiences and possibly anxieties and I think this helped build capacity and people, having seen more of these protocols, now they have a better understanding and now we have several teams that are now able to do this challenge study. [Scientist, Africa]

3.2.4 Potential Individual Benefits of Participation in Endemic Settings

Being infected with a pathogen during HCS often entails few, if any, benefits to research participants. However, if a person is at high risk of infection with the relevant pathogen in day-to-day life, in some cases being infected in the course of research will (i) entail less risk than being infected 'in the wild' (e.g., because of more immediate diagnosis and comprehensive medical care) and (ii) confer a benefit in terms of immunity (whether partial or complete/'sterile') similar to that of vaccination (albeit achieved with a comparatively higher risk intervention) (Selgelid and Jamrozik 2018), that will reduce the risk and/or severity of future bouts of infection (Herrington et al. 1990). Such considerations of individual benefit have not been widely discussed in the ethics literature, perhaps because HCS have, for the most part, taken place in HICs with pathogens that are not locally endemic and/or confer low risks of severe disease. In a recent exception, the 2017 Report on Ethical Considerations for Zika Virus Challenge Trials does mention possible benefits of this kind for challenge study participants recruited in endemic-regions during periods of significant transmission (Shah et al. 2017), and the possibility of such benefits was also noted by Michael Selgelid in a presentation at the 2013 Wellcome Trust Scientific Conference on Controlled Human Infection Studies in the Development of Vaccines and Therapeutics (Selgelid 2013).

HCS might also entail benefits for individual participants in endemic settings if they involve the testing of a vaccine candidate that provides protective immunity against wild-type infection (although the demonstration of protection—or the lack thereof—may be one of the goals of the study and thus be uncertain at the time of enrolment). Such benefits would not arise from the challenge infection itself (and would, for example, also arise in vaccine field trials in endemic settings), yet they may nevertheless (in at least some cases) provide an additional ethical reason in favour of conducting HCS in endemic, rather than non-endemic, settings.

Still, most HCS to date impose a *net* risk on participants (whether or not there are any direct benefits), and it would be unusual if infection as part of a challenge study entailed an expected *net* benefit. Exposure to attenuated pathogens is a vaccination strategy that has been used for some diseases (e.g., live attenuated vaccines for measles, yellow fever) and is being explored, for example, for malaria. However, the benefits of attenuated malaria challenge in endemic settings are the subject of on-going research and hence as yet uncertain (Arévalo-Herrera et al. 2016; Olotu et al. 2018). Thus, although HCS participation in an endemic setting may be less risky and/or more beneficial than participation in a non-endemic setting, participants would still (usually) be accepting a net risk in order to contribute to a research program, the main goal of which is to lead to future public health benefits (as opposed to immediate direct benefits to participants). As Prof. Jonathan Kimmelman suggested in an interview for this project, if we had adequate confidence that people would actually benefit from infection, then we should arguably institute a public health program (rather than a research study) that deliberately infects people with the pathogen in question. Other interviewees did acknowledge a potential for individual benefit in endemic settings that would not occur in non-endemic settings, although it was usually seen as a relatively minor consideration, and one that was contingent on there being a high background risk of infection in the local community (see Box 3.10).

> **Box 3.10 Benefits of HCS participation in endemic settings**
>
> [The] rationale for participating in research is that … you may help yourself and you may help your community and you may help the world and if you do it in … a non-endemic country, then it's just the last one of those; whereas here it's probably all three because, maybe, there is a small chance that an individual volunteering here for a[n HCS] may benefit, in terms of enhanced immunity … when they go back into their malaria endemic homes. [Scientist, Asia]
>
> [Whether an individual participant can be said to benefit] depends on the attack rate where you are and what the probability is [of being infected in daily life, as compared with participating in HCS]. [I]f you [participate in a] challenge [study] you've got a definite risk of infection and an unknown risk of severe complications. [Scientist, UK/Europe]

3.3 Burdens for Participants

Research participation can involve a range of burdens of varying significance, and some HCS designs could, overall, entail relatively high levels of burdens for participants. In research ethics, distinctions are sometimes drawn between risks and other types of burdens, although it has been argued that risks and other burdens should be considered together since, despite apparent differences, they are both (sets of) adverse consequences in the lives of study participants (Rid 2014). On the other hand, study participants may distinguish between what they take to be risks and burdens in various ways (Kraft et al. 2019). Here, we will bracket these debates and use 'burdens' to capture all adverse aspects of research for participants—including exposure to risk (and thus sometimes harm), privacy infringements, restrictions of freedom of movement (e.g., being isolated in an inpatient unit), and/or other reductions in well-being etc.

3.3.1 Limits to Risk

Among other burdens, research often involves risks to participants, although infectious disease research (in particular) also sometimes involves potential risks to third parties (discussed in the next section). HCS may involve varying degrees of risk to participants depending on the study design. Level of risk may depend upon decisions regarding pathogen/challenge strain, study population, whether participants will be inpatients (with close monitoring) or outpatients (with less monitoring), etc. Risks consist of two components: the probability of a harm occurring, and the magnitude of that potential harm (Rid 2014). Furthermore, the magnitude of a harm is a function of its severity and average duration (with more severe and longer duration harms being of higher magnitude and thus more concerning—See Sect. 3.3.4).

HCS have been identified as a group of studies that can, at least sometimes, pose significantly more than minimal risks to participants (Miller 2003), raising questions regarding the upper limit of permissible risk imposition in research among healthy volunteers, and in HCS in particular. Reviews of recent challenge studies have found no deaths or lasting harms among participants (Roestenberg et al. 2012; Darton et al. 2015), however accurate estimates of the risks associated with a particular challenge study (or programme of studies) may not always be explicitly quantified.

As a comparator in healthy volunteer research, one review of phase I non-oncology drug trials performed by one pharmaceutical company found a risk of serious adverse events (those that result in hospital admission, persistent or major disability, life threatening event, birth defect, or death) of up to 0.3% (with no deaths) and a rate of severe adverse events (those that interfere in a major way with a participant's basic daily functioning) of up to 1%; a second review of publicly available phase I trial data found similarly that severe and serious adverse effects comprised less than 1% of all adverse events (Emanuel et al. 2015; Johnson et al. 2016). However, these reviews did capture all phase I (or similar) studies. Rare cases of severe harm among healthy volunteers have included multi-organ failure requiring intensive care admission (in a phase I immunotherapy drug trial) (Goodyear 2006), brain damage and death (in a phase I neurological drug trial) (Moore 2016), and respiratory distress leading to death (in a chemical challenge study where healthy volunteers inhaled an active agent to simulate the pathophysiology of asthma) (Moore 2016).

HCS taken together could be associated with a range of risks, from very low (e.g., studies with low virulence pathogen and/or inpatient studies of treatable pathogens with very early diagnosis and treatment) to significantly higher risks (e.g., infection of immune-naïve individuals in outpatient studies with highly virulent pathogen). Lower risk HCS (e.g., many study designs in current use) might well be safer than many phase 1 drug trials, whereas higher risk challenge designs (and first-in-human HCS) might expose volunteers to significantly greater risk and/or uncertainty. Two questions arise: firstly, should very low risk HCS be classified as minimal risk research?; secondly, if more than minimal risk HCS are in-principle ethically acceptable, what should be the upper limit of risk to which HCS volunteers should be exposed?

3.3.1.1 Minimal Risk

While it is sometimes held that research with healthy volunteers (including HCS) should entail no more than 'minimal risk', or a minor increase above minimal risk, the definition of minimal risk is contentious, and (as above) at least some challenge designs would plausibly exceed such a threshold (as drawn by common definitions) (Hope and McMillan 2004; Resnik 2005). On the one hand, if 'minimal risk' is defined as no more than the usual risks encountered in daily life, then this fails to take into account the fact that some people regularly encounter higher levels of risk in daily life than others, either due to their appetite for high risk activities (e.g., motorcycle riding) or because of the prevailing conditions where they live (e.g., being exposed to endemic infectious diseases, or not having access to adequate sanitation—see Sect. 3.3.3) (Hope and McMillan 2004; Resnik 2005; Wendler 2005; Wendler and Emanuel 2005; London 2006; Shaw 2014). Likewise, setting the standard at the level of the risk that a risk-averse person would encounter in daily life is arguably too strict (Hope and McMillan 2004), and would certainly rule out many HCS—since a risk-averse person would not usually deliberately infect herself with a pathogen. Anecdotally, many individuals volunteering for HCS have a significant appetite for risk. As one scientist describes them: "These were the same young people who would go down the hairiest parts of rivers on rafts" (Cohen 2016).

Alternatively, minimal risk might be specified numerically, for example minimal risk research might be limited to studies involving a less than 1 in a million chance of lasting harm (which would exclude at least some HCS designs—see Sect. 3.3.4) or no more than a 1 in 1000 chance of severe adverse effects to participants.

3.3.1.2 Upper Limit to Risk

Propositions for an upper limit to acceptable risk in research with healthy volunteers include (i) no limit (since consenting adults should be able to decide on the level of risk they will accept) (Shaw 2014), (ii) the risks of high-risk, socially beneficial occupations like fire-fighting (since, similar to research, such jobs involve some individuals taking on net risk to benefit society) (London 2007), or (iii) the risks of kidney donation (since norms in non-research contexts permit this to be done from a sense of altruism, which may also motivate some research participants) (Miller and Joffe 2009).

There is, in any case, no universally agreed upon upper limit for the degree of research risk permissible in HCS. Some argue that it is more justifiable to pursue higher risk research where there is a high likelihood that a given study will produce significant benefits—but most commentators agree that easily imaginable HCS with very high risks[5] would not be justifiable (Hope and McMillan 2004; Miller and Rosenstein 2008; Miller and Joffe 2009). In part, this is for pragmatic reasons, since the public reaction to a case of severe harm in high risk HCS research could lead to a moratorium on other lower risk but potentially beneficial studies (Hope and McMillan 2004). Thus, the need to avoid such an outcome provides (additional) reasons for (i) community engagement around HCS, (ii) careful review of higher risk studies and/or HCS in general, and (iii) enhanced safety monitoring practices during the conduct of HCS (Hope and McMillan 2004; UK Academy of Medical Sciences 2005; Bambery et al. 2015).

---

[5]Consider HCS involving HIV for example.

Many interviewees agreed that an absolute upper limit to risk for would be difficult to define, and some suggested that limits in particular contexts should be partly defined through community consultation, with this in turn being partly with a view to maintaining public trust in research—meaning that even if a higher risk study were judged acceptable under a given ethics framework (e.g., because of high expected benefits), there might be good reasons to find out whether relevant communities would actually accept such a study design (e.g., through community engagement activities) (see Box 3.11).

Box 3.11 Limits to risk and public acceptability of research

[T]he way I understand [limits to risk in the context of ethics review] … is that … we're not just trying to make sure that this specific study is done well and it's ethical – we're also, to some extent, trying to protect the institution of research. [Ethicist, North America]

I also think that those limits ought to be dictated by public perception, to a certain degree … [I]t's not merely a question … [of] how much can we ask an individual to put their lives at stake; it also really bears on how much is the public willing to view this as a kind of legitimate and sanctioned activity, if we put people at this level of risk. [Jonathan Kimmelman, ethicist, Canada]

[W]hat we've learned in my [African] setting, and this is also looking back at some of the studies done here (which we thought were very safe) and how they became problematic … [W]e've learnt that we don't take anything for granted … in the community. We just have to be very careful about it, because it's got the potential to be misunderstood … in all different ways. It doesn't matter whether it is the most safe procedure you thought you were introducing; as long as it is unfamiliar in the community, it is likely to flare up all kinds of rumors. [Scientist, Africa]

[O]n the essential question of 'Can the study proceed from a regulatory perspective?', what are we looking for in terms of ensuring it's safe? And the criteria that we hold to is 'Are there unreasonable risks?' And admittedly that is somewhat of a judgment call about what's reasonable and unreasonable, but the regulatory bar is, is 'Are there unreasonable risks?' [Regulatory representative]

3.3.2 Minimising Risks

It is widely held that risks to participants should be minimised. A key consideration is whether exposing participants to a given risk is necessary to answering an important research question. If not, then such (unnecessary) risks should arguably be eliminated, thus minimising the quantum of risk for a given expected social benefit (US Department of Health and Human Services 1979; Savulescu 1998; Miller and Grady 2001; Hope and McMillan 2004; World Medical Association 2008; Rid et al. 2010; Bambery et al. 2015).

Minimisation of risks to participants during HCS might involve, *inter alia* (i) selection of study populations at lower risk of severe disease, (ii) use of pathogens and/or strains that produce less severe disease, (iii) early diagnosis and/or treatment, (iv) keeping participants as 'inpatients' to enable particularly close monitoring for at least part of the study, (v) close monitoring of (any) outpatient participants, and (vi) careful follow-up for long term outcomes of infection and treatment. Risk minimisation might be considered particularly important in the context of HCS; as one North American scientist argued:

> [Y]ou're completely responsible [from] the moment you … give that injection until that person is clear, either doesn't get it and you've documented that or you've diagnosed it and treated it. So that's different [from] other … non-therapeutic interventions … you have total responsibility and, if you can't be certain that you can take that total responsibility, you have no business doing this. [North American Scientist]

3.3.2.1   Early Diagnosis and Treatment

For HCS with pathogens (such as malaria) where effective treatments are already available, the diagnostic strategy used in a given HCS may significantly influence the risk of symptoms and/or disease as well as the need for inpatient isolation. The potential to reduce such risks will often be particularly important where curative (as opposed to merely supportive) treatments are available. As one North American scientist stated:

> [D]iagnostics [drive] the whole study design in malaria – because if you can't diagnose [participants] until they're pretty sick, then you have to keep them in a hotel. And, if you diagnose them really early, then all that hotel stuff … gets eliminated. And so … we're almost not interested in doing any studies that require hotel phases for malaria anymore because … you don't need to have a whole wing of a hotel when you could just have a more sensitive test. [North American Scientist]

Such observations reveal the ways that study design involves practical and ethical trade-offs. While inpatient studies (and the especially close monitoring these permit) reduce risks for participants, for example, they also increase the burdens related to isolation and close monitoring. Early diagnosis can mitigate both risks and other burdens. Thus, except in situations where there is a strong rationale in terms of the research question that requires delaying treatment, there is arguably a strong ethical case for early diagnosis and curative treatment during HCS for pathogens for which such treatments exist, although thresholds for treatment initiation remain controversial (see Box 3.12).

Even the use of very sensitive microbiological tests and early effective treatment, however, does not always preclude the development of significant symptoms during HCS. One reason that symptoms might occur despite the use of very sensitive diagnostic testing is that some individuals develop symptoms at lower levels of pathogen concentration in the body than others. Unless the threshold for treatment selected as a part of the design of the study is set at or below the level at which the most "sensitive" individual develops symptoms (which might be unknown or difficult to estimate with certainty prior to the conduct of such studies) some individuals may develop symptoms whereas others may not. Investigators may be able to set very low thresholds for treatment in cases where this would not compromise the scientific rationale of the study. In other cases there may be (ethical and scientific) trade-offs between the burdens to participants (experiencing symptoms) and additional important scientific information gained by allowing infections to continue past the point where symptoms develop. Such additional burdens would presumably only be ethically justifiable where similar scientific information could not be gleaned by less burdensome methods (see Box 3.12). For malaria in particular it may be possible to employ treatment(s) against the form of the pathogen that causes symptoms (thus reducing the burdens associated with symptoms among participants) while allowing progression of, and observing, development of other forms of the pathogen (see Sect. 3.4.1).

A second reason that symptoms might occur despite the availability of highly sensitive tests is that there may be trade-offs between a higher frequency of testing (so as to diagnose an infection as early as possible) and the burdens that such close monitoring entails for participants. Since diagnostic tests cannot usually be administered continuously (e.g., 24 hours a day), and since participants have an interest in retaining some privacy and freedom of movement (whether in the context of inpatient or outpatient HCS), the availability of highly sensitive tests and a low threshold for the initiation of treatment might not prevent symptoms from developing in all cases, since there will be a certain amount of time between administrations of the diagnostic test and/or a certain amount of distance between participants and the testing centre (in outpatient studies)—see also Box 3.18 (in Sect. 3.3.7.2) for an example of the development of symptoms in an outpatient.

Box 3.12 Early diagnosis and treatment during HCS

[Some HCS designs are] pushing the boundaries. I know [some scientists] would make [a] case for why they would want to see the severe symptoms in the participants. That is where I would draw some lines and say at the end of the day, in as much as it's about science and new knowledge and benefit for everyone else, but how much are we making people bear the burden and the risk. Is there a level to which we find we are crossing a very thin line between what is ethical and what is not? What is allowable and what is not? And for me severe symptoms are crossing that line very quickly. [Scientist, Africa]

I think it's really inappropriate [to delay] treatment … [S]ome people said … if we delay it for another forty-eight hours then … we'll get more interesting data and we can plot pretty curves of the PCR quantitatively and this kind of stuff. And, for drugs, the longer you leave it the better. I'm very uneasy about that because it seems to me that scientifically, the longer you leave it … even for one hour [or if] you leave it for longer after you confirm infection, there has to be an increased risk of some complication and we know you can get severe malaria. [Scientist, UK/Europe]

> So there's like two camps and everything in-between. The one camp is: we need all the data. We should just let them go all the way to become blood-smear positive [with relatively high burden malaria infection], then we can do PCR on everything and get beautiful curves, and it'll tell us everything. The other camp is: any of these suckers in the blood means it failed so we should treat at the first sign of smoke, right. And doing either extreme isn't great. [Scientist, North America]

3.3.3 Risks to Participants in Endemic Settings

Conducting HCS in LMICs might influence the potential risks to participants in several ways. On the one hand, where local health and related infrastructure is fragile, HCS participants may encounter higher risks due to delays to treatment during outpatient studies if/when they develop symptoms of the challenge infection. It may sometimes be possible to mitigate risks such as these via capacity building of treatment centres and/or inpatient study designs.

On the other hand, HCS in LMICs could involve lower risks of severe disease if they recruit individuals who have (partial) acquired immunity due to prior infection and/or innate forms of resistance to particular pathogens—e.g., genetic conditions affecting red blood cells such as sickle cell that reduce the severity of malaria, an effect demonstrated in one LMIC HCS reviewed below (Lell et al. 2017). It may also be ethically important to purposefully recruit individuals with such traits for HCS that involve testing interventions that (if licensed) would be intended for use in such (sub-)populations, since the safety and efficacy of a given intervention may be different in certain groups.

Furthermore, where participants in a challenge study are at risk of being infected with a pathogen in daily life (e.g., because they live in an endemic area[6]), one might think that, in some cases, this background risk reduces the marginal risk an individual would take on by participating in a challenge study[7] (c.f. Sect. 3.2.4). It may thus be more ethically acceptable, from the point of view of balancing the risks and benefits of a study, to enrol those who already face higher background risk (other things being equal). There was widespread agreement among interviewees that such considerations could be ethically relevant in terms of minimising risk in HCS study design, and might often favour of conducting HCS in endemic populations (see Box 3.13). Research ethics literature regarding background risk more generally (Rothman 1982; Robinson and Unruh 2008), however, provides reasons for being wary about the sentiment that risk imposition

---

[6]Importantly, it should not be assumed that anyone living in a country in which a pathogen is being actively transmitted (in part of the country) is at risk of infection on a day-to-day basis (note, for example, that the malaria HCS in Kenya and Colombia reviewed below actually took place in cities in which malaria is not endemic).

[7]With the exception of pathogens such as dengue, for which the sequence of infections with different strains influences the probability of severe disease (see Selgelid and Jamrozik 2018).

on participants might be more acceptable where background levels of risk are higher when/if (i) higher levels of background risk (e.g., in LMICs) themselves reflect injustices and/or (ii) research participation would significantly increase risk to participants who already face high background risks (while it should be kept in mind that *the absolute magnitude of net/marginal risk increase is a key consideration*, independent of background risk magnitude).[8]

Box 3.13 Background risks of infection and risk to participants

[I]t's less ethically difficult in recruiting volunteers [in endemic areas], considering that you're giving someone an infectious disease, to use volunteers drawn from a population that's at risk anyway, rather than a population that would never be at risk, in terms of justifying the balance of risk. [Scientist, Asia]

[I]f you are already exposed, if you're at a greater risk, [the risk] you're being asked to accept as a result of your … participation in the study is lower and the benefit is going to be the same. So the benefit versus risk profile [is better]. [Ethicist, North America]

[T]here are some compelling reasons [to conduct endemic-region HCS], and that's one of them, that … the background prevalence means there is less of a differential … between the [alternative] of not participating and … deliberate exposure. [Jonathan Kimmelman, ethicist, Canada]

[Y]ou get into all of these questions about whose daily life is the right comparator. Is it a local standard? Is it a universal standard? And so that makes me think it's preferable to do the challenge trials in endemic settings because there's an argument that it's actually lower risk in those contexts. [Ethicist, North America]

3.3.4 Long-Term Risks and Lasting Harms

Most commentators have argued that, as a general rule, HCS should involve infectious diseases that are treatable and/or self-limiting (i.e., resolving without treatment) (Miller and Grady 2001; Miller and Rosenstein 2008; Bambery et al. 2015; Roestenberg et al. 2018a). Some have added the criterion (or interpreted

---

[8]Part of the point of (ii) is that those who favour a Rawlsian account of ethics/justice, which requires making the worst off groups of society as well off as possible, might conclude that it is more acceptable to impose higher marginal research risks on well off participants in HICs (with lower background risks) than to impose lower marginal risks on less well off participants in LMICs (with higher background risks)—because we should avoid worsening the situation of those who are already worst off. A second point of (ii) is that if the net/marginal increase of risk resulting from HCS participation is high enough for those who already face high background risks, then HCS may not be justified even if the net/marginal increase in risk for such participants is lower than would have been the case for participants elsewhere: a comparatively lower level of net/marginal risk increase does not entail an acceptable level of net/marginal risk increase (if the lower level of net/marginal risk increase is itself quite high).

'self-limiting' in such a way) that there be "no lasting consequences" (Miller and Rosenstein 2008) or no "irreversible pathology" (Roestenberg et al. 2018a).

Requirements that there be no lasting consequences or harms are potentially important since (i) some infections can (after partial treatment) reactivate from a latent or dormant form (e.g., vivax malaria), (ii) certain pathogens are sometimes more severe on subsequent infection (e.g., dengue), and (iii) some post-infectious syndromes can lead to lasting morbidity even after the acute infection has resolved or been "cured" (e.g., post-infectious irritable bowel syndrome, post-infectious 'reactive' arthritis, Guillain-Barré syndrome, etc.).

Regarding (i), the dormant form of vivax can usually be definitively treated, although certain individuals are at higher risk of adverse effects of treatment and/or treatment failure, and are thus sometimes excluded from vivax HCS (Bennett et al. 2013). Regarding (ii), the risk of severe dengue (either during or after HCS) can be mitigated via careful participant and/or strain selection, although dengue HCS nevertheless require particularly careful study design (Thomas 2013; Mammen et al. 2014; Larsen et al. 2015; Selgelid and Jamrozik 2018). Regarding (iii), while some post-infectious syndromes have known risk factors (see Table 3.1), meaning that individuals known to be at higher risk can be excluded from HCS, such strategies often cannot prevent these outcomes entirely. The fact that lasting harms have not been documented among HCS participants for these pathogens may reflect careful selection practices and/or relatively low numbers of total participants (in whom, by chance, a rare event has not been observed) and/or publication bias (i.e., events that may have occurred might not have been published).

One HIC researcher interviewed was aware of at least one case of presumed post-infectious arthritis occurring after HCS with an enteric pathogen (the case remains confidential and unpublished, consistent with publication bias mentioned above). More generally, interviewees agreed that such lasting harms were particularly concerning and, at a minimum, require (i) careful risk mitigation strategies, (ii) systems to compensate any participants who experience harm, and, in some cases, (iii) long-term follow-up of participants.. However, there was no clear consensus about what level of residual risk of such outcomes, if any, should be considered acceptable. Some experts noted that several pathogens already used in HCS are (rarely) associated with lasting harms (see Box 3.14 and Table 3.1). Of note, many of the LMIC HCS reviewed later in this report were carefully designed to reduce such risks, usually by exclusion of those with known risk factors for such outcomes.

Box 3.14 Lasting and/or irreversible harms

I'm very uncomfortable with the idea that you might leave somebody with irreversible harm when they haven't been given any possible benefit. [Scientist, UK/Europe]

[A] healthy volunteer … has no expectation of incurring Guillain-Barré syndrome or … a major infection and so … the risk is really … quite pronounced in the healthy

volunteer because of the fact that the counterfactual [the risk of not participating] …
is virtually no risk at all. [Jonathan Kimmelman, ethicist, Canada]

Would I give somebody something that [would cause] long-term risks? … I know
that there's been controversy about Zika [which] can cause Guillain Barré syndrome
… And who knows how many? One in a million. One in a hundred-thousand. That's
a risk … that has to be taken into consideration. [Scientist, North America]

[I]f there's irreversible harm, I think most people would say that that's unreasonable
risk … [T]here may be risk of severe injury in some human challenge studies; but, if
the risk is very low – one in a million, one in a hundred-thousand – perhaps then that
might be considered a reasonable risk. But that's where it does get into judgement.
[Regulatory representative]

Table 3.1 Potential long term risks and lasting harms

	Pathogen	Reported in HCS	Mitigation strategy	Strategy used in LMIC HCS
Lasting harm				
Guillain-Barré syndrome	Multiple—campylobacter, influenza, Zika, *Shigella spp.* etc.	No	Early diagnosis and treatment	Unknown
Post-infectious arthritis	Multiple—particularly enteric pathogens incl. *Shigella spp.*	No (possible unreported case)	Exclude those with risk factors (e.g., HLA-B27), early diagnosis and treatment	Yes
Post-infectious irritable bowel syndrome	Enteric pathogens incl. *Shigella spp.*	No	Exclude those with risk factors	Yes
Long term risk				
Relapse	Vivax malaria	Yes	Long follow-up, early diagnosis and treatment, exclude those with risk factors for treatment failure (CYP 2D6)	Yes
Severe dengue (if first infection during HCS)	Dengue	No	Avoid travel to dengue-endemic areas, Vaccination post-HCS	N/A

HLA-B27: human leukocyte antigen B27, a genetic risk factor for post-infectious arthritis. CYP 2D6: Cytochrome P450 2D6, a genetic risk factor for primaquine treatment failure (identified in previous HCS research) leading to vivax relapse. N/A: no dengue HCS have been conducted in LMICs to date

> [S]ome of these infections or some of these infection models [involve] risks that you
> don't know about and you can develop chronic consequences after these infections.
> And so I think ... we need to fully inform [potential participants] that it's not just
> ... this acute infection, that it could lead to something chronically. [Scientist, North
> America]
>
> [M]ost of the models will have their very rare risks, [just] like [natural] infections.
> Like we know many [natural] infections [are associated with] a rare risk of something
> ... exotic occurring, which we rarely see, but ... if [natural] infections have that, then
> challenge models for sure will have that. And obviously you should try to limit that
> as much as possible, but you're not going to get to 100% safety. [Meta Roestenberg,
> the Netherlands]
>
> The borderline cases [ethically speaking] are [firstly, challenge studies] where the
> diseases are serious and people get really sick ... and the second kind is where there
> are these long term effects that aren't entirely predictable. To me those are kind of
> like two features of the borderline cases. [Ethicist, North America]

3.3.5   Uncertainty

Even for pathogens where the natural history of infection is thought to be well
characterised, the fact that HCS allows for a closer examination of pathogenesis
from the moment of infection means that not all risks will be known, particularly at
the beginning of a research program with a novel HCS design (and/or new challenge
strain) when few (or no) people have yet been challenged. Indeed, HCS research has
revealed new and/or unexpected aspects of certain diseases, and these findings have
led to refinement of HCS exclusion criteria as well as further research (see Table 3.2).

Furthermore, HCS trials of new vaccines may also lead to unexpected adverse
effects, given the potential for unexpected interactions between the pathogen, the
vaccine, and the host immune system (for example, vaccines for RSV and dengue
have in some cases been shown to increase the risk of severe disease upon exposure
to infection after vaccination (Acosta et al. 2016; Wilder-Smith et al. 2018)). In
any case, such uncertainties are not unique to HCS. First-in-human trials of new
drugs have sometimes lead to severe harms among healthy volunteers, such as the
infamous first-in-human phase 1 TGN1412 immunotherapy trial that lead to several
healthy participants being admitted to intensive care (Kenter and Cohen 2006). One
interviewee for the current project specifically cited the TGN1412 trial and made an
analogy with first-in-human challenge studies on this point:

> [W]hen you put something into a human being for the first time you really don't know what's
> going to happen, right? You could kill them. ... So you just have a pilot individual or two
> ... You put in a low dose. You have very close observation. You've done toxicology studies.
> I mean it's so gingerly done. And then you build your way up. [Scientist, North America]

Such uncertainties have implications for consent of HCS participants and ethical
review of HCS, particularly new designs (or designs that have as yet enrolled only

Table 3.2 New and/or unexpected findings in challenge studies

Pathogen	Finding	Reference(s)	Impact
Falciparum malaria	Myocarditis/myocardial infarction with non-obstructive coronary arteries	Nieman et al. (2009), van Meer et al. (2014)	Individuals with cardiac risk factors excluded from malaria HCS
Vivax malaria	Pharmacogenetic factors (CYP 2D6) influencing risk of relapse	Bennett et al. (2013)	Further research regarding this polymorphism; potential exclusion of carriers from vivax HCS; public health implications
Dengue	Serositis (a supposed marker of disease severity) in asymptomatic first infections	Mammen et al. (2014)	Further research regarding dengue pathogenesis

small numbers of participants). As a given challenge strain infection and/or HCS design has been conducted in increasing numbers of participants, researchers (as well as reviewers and regulators) can have greater confidence that estimated risks are increasingly accurate and uncertainty reduced. However, since one of the virtues of HCS is that they involve fewer participants than field trials, larger studies (and/or post marketing surveillance of vaccines) may still reveal rare adverse effects (e.g., interactions between the vaccine, the pathogen, and the human host) that were not observed in the small number of HCS participants in earlier phase studies.

In the 2017 NIH Report on Zika HCS, uncertainties regarding the risks to participants and third parties (e.g., due to sexual transmission) were appealed to as considerations (among others) against conducting of Zika HCS (Shah et al. 2017). However, as better estimates of relevant risks are now available and greater understanding of the underlying mechanisms may allow for mitigation strategies, these considerations might now be considered less weighty (by some). More generally, first-in-human HCS in particular will always involve a significant degree of uncertainty. Although this does not preclude novel HCS, it does justify an especially thorough review of prior evidence regarding the pathogen in question (see Box 3.15).

Box 3.15 Uncertainty

I think with most [HCS designs] you know if you do your screening properly, you hope that nothing will go wrong, but here are people who have conditions that are unrecognized and unrecognizable by the screening tools that we use. For example, this isn't a challenge study, but, for the flu vaccine and narcolepsy, who would've known to screen for HLA-type before giving a flu vaccine, right? So I don't think you could predict everything. [Gagandeep Kang, scientist, India]

I think there are types of challenge studies or models that … are completely safe if not more safe than other types of trials we do all the time. I think what's hard is when we don't have a good sense of the disease and what all of its longer-term effects might be. With those trials it's very difficult to even evaluate what the level of risk would be. And that's where it's not clear to me whether their level of risk is fully in-line with what we already permit. [Ethicist, North America]

[T]housands of people have been in [malaria] studies over the years, and they must have been a lot more nerve-racking [back when] less than a 100 people [had] been in these studies. [Scientist, North America]

3.3.6 Other Burdens for Participants

Almost all clinical research entails multiple burdens (other than risks) for participants. These can range from minor burdens, such as filling out a short questionnaire or being subject to standard medical examinations, to potentially more significant burdens such as the privacy infringement of revealing one's medical or personal information, to major burdens such as a long duration of hospital stay and/or isolation. Certain study interventions will be more burdensome for some individuals/populations than for others. Due to cultural beliefs regarding the value/importance of blood, for example, blood draws (especially of larger volumes) may be especially worrisome, and thus burdensome, to research participants in sub-Saharan Africa, as compared with other groups who may be less concerned regarding blood draws (Saethre and Stadler 2013; Njue et al. 2018).

Since HCS frequently involve multiple study visits, blood draws, and monitoring by study staff—and since inpatient HCS in particular involve significant time away from normal activities—many HCS designs are potentially associated with a level of burdens that would be high compared with most studies conducted in healthy volunteers. There are other analogous cases in non-HCS research, although infrequent, such as metabolic chamber studies (in which participants sometimes spend days in a tightly controlled, isolated environment, for precise measurement of metabolic parameters), which, when they are conducted in HICs, frequently attract high levels of payment (see Box 3.16). Likewise, major burdens for participants in HIC HCS typically attract significant payment of participants; and this has sometimes been the case in LMIC HCS, often because payment for one night in an inpatient facility is indexed to local wages and participants spend 1–4 weeks of confinement during the study (see Sect. 3.6).

Recent social science work with participants has suggested that HCS participation, particularly in inpatient studies, can lead to a wide range of burdens and/or secondary effects on the families of participants. In one LMIC study, for example, the children of participants were unable to attend school while a parent was participating in HCS

(Njue et al. 2018). One HIC study found that outpatient HCS participants often encountered significant disruptions to their daily lives as a result of participation (Kraft et al. 2019). Many stakeholders we interviewed for this project felt that the burdens of participation (i.e., burdens other than risk imposition) raised ethically important issues for HCS design in need of further analysis.

Box 3.16 Burdens of participation

[HCS protocols often] keep people in residence for a long period of time. I think that's pretty unique. I don't think we do that for many other studies ... just that phenomenon of saying to people – you know, you might need to be in residence for a month or even longer, six weeks, you'll need to stay here and you will not be able to leave, under any circumstances ... I don't think we fully understand what the ethical implications of that are. [Scientist, Africa]

[T]here's physical risk, which I think for [some HCS] is quite small, but there is also the emotional risk ... but the bigger thing is the burdens. [In some HCS designs] you have to be in residence for fourteen days, minimum, [and] being in residence means that you have to make sure that other parts of your life ... and kids, and jobs [are taken care of] ... so that's quite a big commitment, and a sacrifice, I would say. [Scientist, Asia]

How can we make [participation a good] experience good for them? What kind [of] residence would that be required to be? ... If we are curtailing their freedoms of movement, how does that then balance against the risk? ... We are telling them not to go to endemic areas, even when they do get out of there – they're still within the study. In other words, we're interfering with their freedom, and how do we take account for that? [Scientist, Africa]

[Y]ou pay a lot of money for someone to stay as an inpatient. [For example, in a] metabolic chamber study ... people were getting paid $6,000 for that, because you're in a chamber for a month. [Ethicist, North America]

3.3.6.1 Mental Health

Since challenge studies often involve significant burdens for participants, these burdens could plausibly include and/or lead to deterioration in mental health (whether or not the participant had a prior history of psychiatric illness). Such risks may be particularly significant during prolonged inpatient studies involving social isolation, and some research groups have adopted careful psychological screening of potential participants to inform judgements about their ability to tolerate periods of isolation (sometimes of several weeks' duration) (Pitisuttithum 2018). Even with careful recruitment practices, one previous LMIC HCS recorded a serious adverse effect related to a participant who was briefly admitted to hospital due to an anxiety crisis (by comparison, in the same study, no physical serious adverse events occurred from the challenge infection) (Herrera et al. 2011). The mental health of participants is thus potentially an area warranting further social science work and/or ethical analysis; the exclusion of any potential participant with a psychiatric history may be overly restrictive and/or unfair, thus a nuanced approach is needed. As one stakeholder noted:

> [In] Oxford we have a very high proportion of volunteers who have a known psychiatric diagnosis: anxiety, depressive. The majority of those are managed ... it doesn't affect their activities of daily living. As to working, maybe they take antidepressants. So it's not as simple as just saying 'Anyone with any psychiatric history is not suitable.' [Scientist, UK/Europe]

3.3.7 Participant Behaviour

All research with human participants involves the possibility of unexpected human behaviours. In the context of HCS, certain participant behaviours might lead to greater risks than those anticipated in the study protocol—for example, participants (i) choosing to withdraw from the study and/or refusing to be treated after challenge infection, and/or (ii) leaving the study site and/or becoming uncontactable after being infected. Since researchers are exposing healthy volunteers (and sometimes third parties) to potential severe harms (e.g., if an infection with malaria were to go untreated), investigators arguably have especially weighty ethical responsibilities to ensure that the risks entailed by such human behaviours are minimised. It is perhaps also the case that participants who consent to be infected have an ethical responsibility to abide by monitoring, treatment, and/or social distancing requirements during and/or after the study (especially where not complying might entail risks being imposed on others). We discuss the right to withdraw and the risk of participants absconding (i.e., leaving the study without informing research staff) below.

3.3.7.1 Participants' Right to Withdraw

The right of participants to withdraw from a study—at any time, for any reason, and without having to give a reason—is widely endorsed in theoretical research ethics and in practice, although the practical implications of this right have been a matter of debate in particular contexts (Edwards 2005; Helgesson and Johnsson 2005; McConnell 2010; Schaefer and Wertheimer 2010). If a challenge study participant were to exercise this right after being infected with certain pathogens and before the infection has resolved (with or without curative treatment) then this might increase risks to participants themselves and/or risks to third parties. Many HCS researchers interviewed for this project recognised the difficulties that might arise in such contexts and the potential for adverse outcomes that could undermine public trust in research, including (in some cases) a higher potential for third-party risk in LMICs. Investigators have tried to account for this in study protocols;

however it was seen as an unresolved issue in need of further analysis and guidance for researchers (see Box 3.17).

Box 3.17 Participants' right to withdraw

[T]hese are adults and … their participation in the study is voluntary. They could always withdraw their informed consent. They could walk even though that is a risk to the community at large. We can't hold them against their will. And that was a concern. And so we spent a lot of time emphasising to them … early on in the trial, how important it was that they complete the treatment and they complete the follow-up. And that we understand that it's a long time. [Carl Mason, scientist, USA]

[I]n our consent forms, we [advise participants that] in the event that you wish to leave the study, of course you're allowed to do this, but we would expect you to complete a course of treatment … But I think if there was a scenario where somebody left and they wouldn't take the treatment, we didn't have a protocol to follow in that scenario. I think we would have to let them go. We did say if somebody went missing … we would notify the local authorities to search for them because there would be a concern about their mental and physical health. And we would contact their next of kin to try and locate them … Of course, in the UK, the concern [in malaria HCS] is not about transmission, it's about wellbeing of the patient. But, you know, in Nairobi [given the nearby presence of vector mosquitoes] there's a potential to have onward transmission as a result of not having treatment. [Scientist, UK/Europe]

Another issue I see as more complicated in the context of challenge trials is the right to withdraw … [W]ithin the context of the US regulations it's considered a right that people have, that they can exercise at any time. There are other trials where you still can't quite just leave when you want to leave because it may not be safe for you, but challenge trials are trials where that issue becomes difficult. And I think it would be helpful to know more, to have a better public health framework maybe similar to what we think about for quarantine, to understand what [are] the limits of measures to restrict someone's liberty, if they're in a challenge trial, like when is that even acceptable and to what extent. [Ethicist, North America]

[W]hat we have in these consent forms [in Gabon], and what we also explained to them, is that they can leave anytime – but when they have been infected … they need to come back to be treated. [Benjamin Mordmüller, scientist, Germany]

[T]here are reviews of the data on informed consent [but] the right to withdraw is less well-understood in low-income countries. And if that's right the rates of withdrawal are something that is really tricky in the context of challenge studies because there may be times when it's not safe for someone to leave the research. [Ethicist, North America]

3.3.7.2   Risk of Absconding from Studies

Participants have occasionally absconded from HCS after being challenged, including one high profile case of a participant in a UK malaria HCS who was eventually located in the Netherlands.[9] Absconding from a study might in some scenarios be a special case of the right to withdraw (i.e., participants exercising the right without notifying study staff), or, in others, it might reflect impaired decision-making by a participant due to physical or mental illness. Many of the scientists we interviewed identified the possibility of such events as a significant concern that had (in some cases) led to revisions of study design and procedures (see Box 3.18). One solution might be closer monitoring of participants. However, the more closely participants are monitored during the study (whether on an inpatient or outpatient basis), the more this monitoring may be burdensome for them. Thus there are important ethical trade-offs to be made in study design between post-challenge monitoring that is sufficiently close to minimise risk but not so intrusive as to be overly burdensome.

Box 3.18 Risk of participants absconding and risk mitigation strategies

[W]e had someone in one of our typhoid studies who absconded because he was an actor and had an audition, which he hadn't been expecting, for a lead role in a play ... [W]hilst he was developing typhoid he went to do his audition, and we lost touch with him and we were very worried about him and in fact he got the role ... and he actually had positive blood cultures for typhoid at the time. [Scientist, UK/Europe]

[W]e had one volunteer leave from one of our studies ... [H]e'd been challenged as part of that study and he decided he'd go and see his uncle [in another country]. And fortunately we caught him and we made sure he took [treatment for his infection], and so on. [Scientist, North America]

There was talk of once in a while somebody sneaking out and going to the shops but not much of like, somebody being away in terms of going home and putting others at risk ... Somebody will sneak out of the gate of the university and go to the shops nearby and come back. [Scientist, Africa]

[A]s a result of our lost volunteer ... we wrote into our consent document and our volunteer information sheet and our protocol, actions that we would undertake if a volunteer went missing. [Scientist, UK/Europe]

3.4 Risks to Third Parties

Many types of research with infectious diseases can pose risks of infection to those not directly participating in the research (i.e., third parties) (Kimmelman 2005; Battin et al. 2008; Eyal et al. 2018; Shah et al. 2018). In the case of HCS, the principle risks to third parties are related to transmission of the challenge strain(s) from infected participants to others—and potentially onwards to many more people. Third-party

---

[9]https://www.theguardian.com/society/2010/oct/19/malaria-trial-nurse-found [Accessed 29 March 2019].

risks are sometimes referred to as 'bystander risks'; we eschew this term here, because an infectious disease can, via a chain of transmission (sometimes of great distance and/or duration), harm distant others, not only bystanders (i.e., those who happen to be present when the research is taking place although not themselves participants).[10] Risks to third parties include the possibility of transmission (and/or harm) to unborn children, providing a reason to exclude pregnant women from HCS (Shah et al. 2017). Other third parties at risk can include participants' family members, but also members of the general public (Miller and Grady 2001; Hope and McMillan 2004; Bambery et al. 2015; Shah et al. 2017).

Other effects on third parties may be relevant for certain HCS designs. Where live-attenuated vaccine strains are tested in HCS (particularly vaccines for enteric pathogens such as *Shigella*), for example, there is sometimes a small probability of transmission of the vaccine strain to third parties (Kimmelman 2005) (although, given adequate attenuation, the potential harms are low, at least for immunocompetent individuals, and there may even be a net benefit of such transmission (Paul 2004)). Where challenge with vector-borne pathogens is administered by mosquito bite, furthermore, the potential introduction of a new vector species (imported from elsewhere for the purposes of HCS) could alter local ecology[11] (were the vectors to escape from the study facility) and, in endemic settings, the epidemiology of vector-borne diseases transmitted by this vector (Orjuela-Sanchez et al. 2018).

Some have argued that researchers have extensive ethical duties to third parties where there is a risk that infection will be transmitted to them from research participants. In the context of infectious disease research more generally (though not focused specifically on HCS), Battin et al. have argued that, where it is possible to identify specific people at risk, researchers should obtain individual informed consent from these third parties before commencing a study and/or, where the risks are significant but particular third parties are not readily identifiable, some form of community consent should be sought (Battin et al. 2008).

One way to obviate the need for such additional consent procedures is to reduce third-party risks to near zero by (i) rigorous infection control and biosafety procedures at HCS research centres, and, in some cases (ii) strict isolation of participants (e.g., by keeping them in an 'inpatient' setting for the period in which they are potentially contagious), although this in turn entails significant burdens for participants. Alternatively, where such third-party risks are not reduced to near zero they could in some cases be monitored and/or quantified by (enhanced) public health surveillance in the local area including genotyping of strains detected to assess the degree to which the challenge strain is transmitted to the local population. For example, one recent paratyphoid HCS design by UK investigators explicitly included provisions for the institution of such public health surveillance

---

[10]Others have opted for 'risks to nonparticipants' (see Eyal et al. 2018).

[11]This risk is not unique to challenge studies, as other types of (vector-borne disease) research sometimes involve maintaining colonies of vector species and, in some cases, the importation of such species.

measures in the event of presumed or suspected transmission of infection from participants, including the provision of the challenge strain to local public health microbiological laboratory for comparison with any clinical isolates (e.g. those detected during a public health outbreak occurring around the time of the proposed study) so that it would be possible to make accurate assessments of whether third-party transmission from study participants had occurred (McCullagh et al. 2015). This, however, might significantly increase study costs and require capacity building of local public health laboratories in LMICs.

In any case, while some may debate whether such strict duties (to obtain consent from any third parties at risk) always apply, the potential risks (even if small) may vary in different contexts depending on the mode of disease transmission. For vector-borne diseases such as malaria, if there are no local vectors then there are minimal risks of third party transmission (apart from blood donation by participants while infected) (Herrera et al. 2009; Hodgson et al. 2014; Hodgson et al. 2015). The risks of transmission of diarrhoeal pathogens via sewerage systems have been considered in reviews of HCS protocols (Cohen 2016; Pitisuttithum 2018). Such risks may be low in HICs with adequate sanitation, but could be higher in communities with poor access to sanitation (e.g., in LMICs), suggesting a strong rationale for inpatient studies and/or robust biosafety procedures in such settings (Pitisuttithum 2018).

Conducting HCS in endemic LMICs may affect the potential third-party risks of such studies in multiple ways. On the one hand third parties with pre-existing immunity (and/or other forms of resistance to the infection in question) may be at lower risk of severe disease were they to be infected as a result of third-party transmission from study participants. On the other hand, other third parties might be at higher risk, including children (especially those who are malnourished, unwell for other reasons, etc.), those with health issues including other chronic infections (e.g. HIV), and/or those with poor access to healthcare. More generally, since populations in LMICs often have higher levels of ill health partly because of inequities in the social determinants of health, some might consider it less ethically acceptable (and/or potentially more unjust) to impose third-party risks of infectious disease in such contexts, even if the background risk of the infection in question is already relatively high[12] (e.g., where HCS are conducted in an endemic area—see related discussion of background risk in Sect. 3.3.3).

Thus, the potential for higher third-party risks in certain contexts can lead to controversial questions. How important, for example, is a small third-party risk and/or single episode of transmission (e.g., from a study participant to a third party) in the context of high local endemic transmission (and/or high average local levels of immunity)? Some individuals and communities may consider this additional risk negligible, while others may see each additional episode of transmission as highly significant –stakeholders interviewed for this project held widely divergent opinions

---

[12] A further relevant consideration, depending on the disease in question, is whether the challenge strain is already prevalent in a given endemic area (since participants and third parties may have immunity from prior infection with locally prevalent strains that would not necessarily protect against severe disease when infected with a different challenge strain).

on this matter (see Box 3.19). Given this potential controversy, and given the potential for third-party risks to undermine public trust in research (see Box 3.20), the potential for such risks would constitute an additional reason for community engagement (to assess community views on the importance of such risks and/or to seek community consent for the research to proceed) and for carefully designed research procedures that reduce transmission risks.

Box 3.19 Third party risk and background risk

I think [the risk to third parties of a malaria HCS in a highly endemic area] is a very, very small risk only if you can even, or should even, call it a risk –[because] eighty per cent and more [of the local population] harbour malaria parasites at any given time point. [Scientist, UK/Europe]

I think it really depends on the background transmission rate … [I]f you're working in a hyper endemic setting, I just don't think there's any quantifiable increased risk to the population … [P]eople are being infected every week, all the time, so … I don't think that's a risk, a real increase[d] risk for the population. [Scientist, UK/Europe]

[If] there is not much greater risk [to third parties, compared to background risk] and you are not using a strain that is resistant to any of the drugs that are available, then people [once they understand this] will be much more comfortable I think … most of the risk that we see are much more academic than real [or] practical. [Scientist, Africa]

[In Thailand, even if a participant were able to leave the isolation ward, local population] antibody levels were much higher than the antibody levels that were seen in volunteers in the previous studies in the US. So for [*Shigella*], and probably for some other diseases as well, such as malaria in challenge studies in endemic areas, you're going to have people [in the general population] that have partial immunity. And the risk [to third parties] might actually be a little less. [Scientist, North America]

[C]ontainment is possible. It's expensive. Not so expensive in developing countries as it is in developed countries, but it's possible and if you can minimise risk [to third parties] you should do so, and remember that it's a drop in the ocean, but it's a drop in the ocean that can result in death. [Scientist, UK/Europe]

I think of the response if an individual is inadvertently infected. A third party individual is going to [feel] different[ly about it] if they later learn that it's because they came into contact with someone who was in a scientific experiment, than if it's [just] because of a mosquito … [R]isk is, or has, these … moral layers … that we all … bracket when we talk about risk … in a quantitative way. [Jonathan Kimmelman, ethicist, Canada]

Box 3.20 Third party risk and public acceptance of research

[T]hird party risks are a very, very important component, both on ... first principle research ethics that ... if there are third party risks there are risks that are being born involuntarily by other individuals but also from the standpoint of public perception ... [W]hen the public perceives a risk that is enduring, [and] as having been involuntarily endured, then there tends to be much more acrimony and controversy than when the public feels that ... there's a voluntariness and an awareness ... and so I think those kind of third-party risks are the kinds of risks that you worry about in terms of destabilising or undermining ... public support for research. [Jonathan Kimmelman, ethicist, Canada]

We just need someone with typhoid on the Oxford study to go and not tell them they're working in a food van and have an outbreak in Oxford and that will be it [for the whole field of challenge studies]. [Scientist, UK/Europe]

3.4.1 Third-Party Risks and Studies of Transmissibility

Some HCS are designed to investigate the transmissibility of the pathogen in question. For example, such studies might involve measuring the number of microbes in the blood and/or stool of an infected participant (the idea being that the number of microbes present is in many cases correlated with the risk of transmission to others). Investigation of transmission was one of the goals of several historic yellow fever and malaria HCS (see Sect. 2.2) and was also identified as an area potentially in need of further work by malaria HCS researchers based in Kenya (Hodgson et al. 2015). HCS that specifically aim to investigate the transmissibility of the challenge infection warrant particularly careful design with regards to third-party risk (e.g., because studies not investigating transmissibility can be designed with early curative treatment, whereas transmissibility studies will often be required to leave participants infected with the challenge strain for longer periods of time, during which they may be able to transmit the infection to others).

Australian investigators recently conducted a falciparum malaria HCS assessing the transmissibility of falciparum malaria in malaria-naïve Australian volunteers who were treated post-challenge so as to reduce symptoms among participants (since malaria symptoms are caused by a particular form of the parasite, amenable to specific treatment) without affecting the transmissible forms of malaria (gametocytes). These transmissible forms were then measured by feeding mosquitoes on the blood of participants so as to provide a model of transmissibility against which transmission-blocking interventions could be tested (Collins et al. 2018). To our knowledge the only similar HCS testing transmissibility in an LMIC is a vivax malaria study in Colombia that was based on secondary analysis of samples from one of the HCS case studies reviewed below (Arévalo-Herrera et al. 2014). Both the Australian and Colombian studies were performed (using mosquitoes in tightly controlled laboratory settings) in non-endemic areas lacking malaria vector mosquitoes in the wild, meaning that there would be effectively no risks to third parties. Were such transmissibility studies to be conducted in malaria-endemic settings and/or areas with local vector mosquitoes, study design and review would arguably require careful assessment and/or mitigation of third-party risks.

3.5 Participant Selection

Since the formalisation of ethical principles for research, there have been debates regarding how fairness should be understood in the context of selecting participants for research and the extent to which special considerations apply to recruitment from vulnerable populations (National Commission for the Proptection of Human Subjects of Biomedicaland Behavioral Research 1978; Meltzer and Childress 2008).

Individuals and populations can be vulnerable in different ways, and the term 'vulnerability' is generally used in research ethics to identify those who are more likely to be exploited and/or harmed as a result of participation in research (Macklin 2003; Luna 2009; Rogers et al. 2012; Lange et al. 2013). Individuals might be physiologically vulnerable (e.g., at higher risk of severe harm as a result of participation in research), socio-economically vulnerable (e.g., perhaps more likely to be influenced by payment for research participation—see Sect. 3.6), and/or vulnerable in terms of being unable to provide autonomous consent (e.g., children—see Sect. 3.5.4) or because of dependence on others (e.g., children, institutionalised individuals) (Goodin 1986; Luna 1997; Macklin 2003).

A particular concern regarding exploitation of vulnerable populations is that relatively privileged populations may be the primary beneficiaries of research conducted in more underprivileged populations (Macklin 2003). However, conducting HCS in LMICs may be less exploitative in this regard if (i) the findings from such studies are thus more relevant to the (LMIC) populations in which the pathogen being studied is endemic and (ii) scientific knowledge and/or new interventions produced by HCS are ultimately made available to those populations (Wenner 2015, 2017).

An increasingly recognised problem is that excluding vulnerable populations from research as a way of protecting them from further burdens can ultimately lead to these same populations being excluded from the benefits of research (Sheffield et al. 2018). It may sometimes thus be ethically important to include vulnerable populations (with appropriate measures to minimise burdens), especially where the results of research in other populations are not likely to be generalisable to the vulnerable populations in question. This is one consideration that is sometimes in favour of conducting (more) HCS in LMICs.

3.5.1   Vulnerable Populations in Human Challenge Studies

Since many pathogens for which HCS might be considered occur at higher rates in poor LMIC communities that are particularly vulnerable (in multiple ways), fair participant selection for HCS may be especially complex (World Health Organization 2017). Concerns regarding the potential for exploitation of, or harm to, individuals from vulnerable populations might thus explain why more HCS haven't been conducted in LMICs to date. However, there are several countervailing considerations. First, in cases where the results of HCS in non-endemic HIC populations are not likely to be generalisable to LMIC populations where a given pathogen is primarily endemic, there may be both scientific and ethical reasons to recruit HCS participants from an endemic population (Hodgson et al. 2015) (see Sect. 3.2.1.1). Second HCS in endemic LMICs might sometimes entail relatively lower risks and/or potential for direct benefits to participants (e.g., related to immunity) as compared with those in non-endemic settings—meaning that HCS in some 'vulnerable' populations might (perhaps counterintuitively) involve less risk to participants (Selgelid 2013; Lell et al. 2017; Selgelid and Jamrozik 2018) (see Sect. 3.3.3). Thirdly, undertaking HCS (and/or other kinds of research) only in HICs might undermine efforts to build research capacity in endemic settings and exacerbate the research neglect of certain pathogens (Selgelid and Jamrozik 2018). Finally, recruitment in HICs might still end up disproportionately selecting vulnerable individuals, especially since HCS often involve large time commitments and/or isolation from other social activities, meaning that, for example, unemployed people and/or students might be overrepresented in HIC study populations (see Box 3.21) (Elliott and Abadie 2008).

Many of those interviewed for this project supported conducting HCS in LMICs because the results may then be more relevant/generalisable to the eventual target population for novel interventions. Some noted that although the rates of poverty and some other vulnerabilities are often higher (on average) in LMICs, LMIC populations should not be labelled as vulnerable *en masse* and thus excluded from research. Each community, in HICs as well as LMICs, includes a range of individuals with different levels of various vulnerabilities. Whether HCS are conducted in HICs or LMICs it is thus important to evaluate the specific vulnerabilities of the (potential) study population and design studies in ways that reduce the chance of harm and/or exploitation accordingly.

Box 3.21 Participant selection and vulnerable populations

I do think location matters. I understand the concern about exploiting populations that are already vulnerable because they're living in an endemic setting, and, like, "Oh well, now you're adding to their risks." But I, I think there are reasons to do it in endemic settings first. [Ethicist, North America]

[O]ccasionally I see people ... neglect the fact that in low income and middle income countries you have islands of affluence, and in high income countries you have islands, if not continents, of disadvantage. [Jonathan Kimmelman, ethicist, Canada]

I'm completely committed to the value of fairness. But I think we, as a result of that, should not act as if persons in low- and middle-income countries were incapable of altruistically participating in research that benefits others. [Ethicist, North America]

[HCS participants are] usually younger people because we need healthy people, and they tend to be people who have enough time to manage this. So ... sometimes it's so demanding in terms of time that people who have high-stress jobs or very busy families can't really participate. And so it's generally people who just really care about contributing something and feel good about contributing something to science and biology. [Scientist, North America]

I wouldn't say all of them but a significant proportion are people that are unemployed, you know. And so ... you're giving them a nice opportunity for income. And whether they understand ... And then that needs to be balanced with them fully understanding the risks. [Scientist, North America]

3.5.2 Consent

Informed consent for research participation involves a potential participant with adequate cognitive capacity who is adequately informed regarding the details of a study, understands this information, and makes a voluntary, uncoerced decision to participate. Among those interviewed, there was widespread agreement that HCS per se did not need a special consent process. However, many supported the notion that (in research in general, and in HCS in particular) the higher the burdens (including risks) of a given study, the greater the responsibility of those conducting the study to ensure that participants are able to give fully informed consent. HCS that involve new or particularly complex models, higher risks, and/or those recruiting individuals with little prior experience and/or understanding of research may thus warrant an especially careful consent process (see Box 3.22). Our review of LMIC HCS case studies found that consent processes in HCS frequently involved multiple information sessions and/or tests of understanding, suggesting that investigators recognise the importance of more stringent consent processes in such studies.

Box 3.22 More stringent consent requirements

[W]hen you have elevated risk ... that expectation for understanding and for voluntariness is much more enhanced than when you have low levels of risk, and so under those circumstances there ought to be a much more demanding and exacting consent process. That really seems straightforward but that also seems embedded in the standard notion of ... how we do informed consent so [it] doesn't seem that claim is a radical departure from what we already believe and practice. [Jonathan Kimmelman, ethicist, Canada]

[T]he greater the risks of study participation, the greater your duty to really be sure that your subjects are fully understanding what's at stake. [Ethicist, North America]

[T]he things that do actually improve understanding are giving people more time to think about a consent document and then testing their understanding, and then re-educating them on the things they don't get right. So those things I think probably should be more universally implemented than we do right now. And they would also be useful for challenge studies. [Ethicist, North America]

CHIM studies might be a good example of a circumstance where you'd actually want to build in some comprehension checks and that's definitely not just an issue for low resource settings [Ethicist, North America]

So [in Gabon] we explained [the risks of the study] but we also have a quiz afterwards, so they actively have to answer these questions by themselves … [W]e have a quiz with multiple choice answers, and they have to pass this quiz [so] that we can somehow have evidence that they really understood. [Benjamin Mordmüller, scientist, Germany]

3.5.3 Education Level

It is sometimes thought that it would be more ethical to recruit those with higher levels of education as research participants because this may improve informed consent (if educated participants more easily understand information about the study). Some HCS (including in LMICs) have thus aimed to recruit tertiary-educated individuals and/or university students (especially medical students) in particular (Hodgson et al. 2014; Shekalaghe et al. 2014). Despite these apparent advantages, there are also several ethical disadvantages of such a recruitment strategy: (i) excluding less well-educated individuals might be unjustified if they are able to understand a study well enough to provide adequate informed consent, (ii) university students (or those who have received university education) may not be representative of the eventual target population for an intervention (e.g., because they are more likely to be affluent and/or to live in cities and less likely to live in highly endemic parts of LMICs and/or because in some countries women are much less likely than men to receive university educated), (iii) excluding less well educated individuals from HCS research may thus be unfair, especially where poor and/or less well educated individuals are at higher risk of the disease in question and/or where women are underrepresented in tertiary education (leading to the exclusion of such individuals from research), (iv) students may feel pressure to participate (e.g., from academics within the faculty with an interest in the study) making consent less voluntary (Bonham and Moreno 2008), (v) educated individuals (e.g., healthcare workers) may sometimes actually be less compliant with study protocols than other potential participants. Two anecdotal cases support the latter point: (1) the participant who absconded from one falciparum malaria HCS (discussed below) was a healthcare worker, and (2) in a Colombian malaria HCS discussed below, one participant who worked as a paramedic was strongly suspected to have self-treated with antimalarials after challenge, thus undermining the scientific value of his or her participation in the experiment (Herrera et al. 2009).

3.5.3.1 Education Level in Low- and Middle-Income Countries

In practice, LMIC investigators have sometimes been successful in recruiting enough tertiary-educated individuals for HCS (Shekalaghe et al. 2014; Jongo et al. 2018), whereas others have found it difficult to recruit as many students as planned (and thus recruited others with lower average education levels (Hodgson et al. 2015)). Social scientists embedded with more recent challenge studies have suggested that many less educated individuals appeared to be able to provide adequate informed consent, especially with well-designed community engagement and multiple opportunities for careful explanation of the study (Njue et al. 2018). Including less educated individuals can help researchers to recruit more people from rural, highly endemic areas, and thus learn more about acquired immunity and the efficacy of interventions in those at particularly high risk of the infection in daily life (Hodgson et al. 2014; Hodgson et al. 2015). More work will be required in other settings to assess the quality of informed consent and whether the presumption in favour of recruiting especially well-educated individuals is justified.

There may be one additional way in which recruiting those who live in or near highly endemic areas, even if they are less educated, is ethically preferable (so long as adequate informed consent is assured): such individuals may be more likely, on average, to have an interest in the goals of the research because prior experience of the infection in question (e.g., in themselves or those close to them) may lead to greater understanding of the need to reduce the harms of such familiar infections and thus motivate participation in research, above and beyond a more general sense of altruism that may motivate individuals in non-endemic areas (see Box 3.23) (London 2005; Njue et al. 2018).

Box 3.23 Consent in LMICs where HCS pathogens are endemic

Some people worry that, if you do a challenge study in an endemic setting, you have to do more education because people would be less likely to understand what the study actually involves. And there … the data on informed consent actually are somewhat reassuring … [T]he data on informed consent suggests that there isn't a systematic difference in terms of geography, in terms of what people understand. [Ethicist, North America]

> [In] a country like Brazil, where Zika has been a problem ... there are people who are willing to ... be soldiers against the disease that they see afflicting ... their peers, as opposed to ... a low income person in Baltimore, who is unlikely to get Zika exposure, unlikely to know someone who has Zika, and is [participating in research] to pay the rent. [Jonathan Kimmelman, ethicist, Canada]
>
> [W]e had [one paticipant] who'd never been to school, but he passed the test of understanding ... and we had a few [with] really the low primary level kind of education ... So, there were all kinds of education levels. [Our work with these participants] helped us to ... realise it does not necessarily have to be the level of education that mattered, it's about understanding what the key elements ... of this study are. [Scientist, Africa]
>
> I wouldn't call [consent practices in LMIC HCS] special ... [E]verywhere we go the informed consent is adapted to the local setting. And some places are better-educated than others. Some have two or three different languages that you ... need to translate into ... The informed consent doesn't change hugely between different sites, and we have had steers in the past by the IRBs to say ... 'You can make this a bit more simplified,' or, 'You could clarify that.' So we rely heavily on the advice of the [LMIC] IRBs to help us with that as well and get some excellent feedback from them in that regard. [Scientist, North America]
>
> If you're developing a vaccine, and you're planning to actually deploy that in super rural areas where the majority of the population is illiterate, obviously you would want to move that controlled human infection model to that population also because that population for lots of scientific reasons might respond differently so you need to ... research whether it's going to work in that population as well. [Meta Roestenberg, the Netherlands]

3.5.4 Children

One of the most controversial questions regarding HCS is when, if ever, it might be ethically acceptable to recruit children as participants in studies that could cause significant symptoms/disease among (child) participants. On the one hand, many pathogens of interest (e.g., falciparum malaria, *Shigella*) predominantly harm young children in endemic settings who would thus stand to benefit most from new vaccines (and thus the research in question). There would thus be a *scientific rationale* for HCS in children. On the other hand, (i) HCS in carefully selected adult volunteers might be sufficiently generalisable to children at risk (and thus obviate the need to recruit children), (ii) children lack the capacity to provide informed consent to participation, which is especially important given that the consent of participants arguably matters more for burdensome and/or higher risk studies (as discussed above), (iii) challenge infection might sometimes be higher risk in children, and (iv) public perceptions of the enrolment of children in HCS could undermine trust in HCS and/or research more generally.

For these and other reasons, the only HCS performed with children have used (live-attenuated) vaccine strains of micro-organisms as the challenge agent (Groome et al.

2017)[13]; recent/modern HCS have thus avoided the use of disease-causing challenge organisms in children (WHO Expert Committee on Biological Standardization 2016; Baay et al. 2018). Among those interviewed for this project, there was widespread consensus that, even if HCS in (involving disease-causing organisms) in children could be conducted safely, there should be a presumption against enrolling children in HCS (involving disease-causing organisms) with HCS in adults and/or field trials in children as plausible alternatives to be tried first. Many suggested that a very carefully described rationale and wide consultation would be required before any such study were considered. A particularly strong theme was the issue of public and/or community acceptance and the risk of undermining trust in research (see Box 3.24).

Box 3.24 HCS in children

[T]o do these challenge studies for some of these diseases that occur so early in life, you know … a child with diarrhoea … in order for you to really get a good readout on the value of vaccine for child diarrhoea, you can't do it in an adult. You have to do it in a child. But … I don't think we're talking about doing human challenges in children, because I think that there's a lot of issues there in terms of consent. [Scientist, North America]

[A]s a researcher, I would say you could do [a malaria HCS in children with early treatment] safely, but getting it past an ethics committee would be a massive challenge. Certainly, you know, giving kids malaria … the optics of it are not good. So, it would have to be preceded by a massive public and stakeholder engagement campaign. [Scientist, Asia]

I think we might be able to justify [a *Shigella* HCS vaccine trial in children] but it would … definitely be on a case-by-case basis and there would have to be tremendous consensus both in the host country probably as well as globally. That's something you would have to take to [a] body like WHO and really try to build a consensus. [Carl Mason, scientist, USA]

[T]here are certain kinds of risks and certain kinds of incidents that tend to, more than other kinds of incidents, galvanize public opposition campaigns and among those I mentioned is involuntary exposure but among those is risk to children. [Jonathan Kimmelman, ethicist, Canada]

[Y]ou know the difficultly of doing vaccine research in children and you just have to have a few things, coincidently, go wrong and you can destroy a whole program of research or public health implementation. [Scientist, UK/Europe]

I think … in general most people would say that it's just … unacceptable to do challenge studies in children. I think that's most people's starting point, and I think before we move away from that position, we'd have to be on really very solid ground … I hope nothing like that proceeds without all sorts of very extensive consultations and discussions. [Scientist, UK/Europe]

---

[13]In such cases, exposure to the live-attenuated vaccine strain as a challenge agent would entail an extremely low risk of harm and perhaps even a net benefit to participants.

[T]here's already, to me, a serious public relations issue about challenging people with a pathogen … and if you were then to translate that to doing that in a group who can't provide consent and something goes wrong I think is obviously a disaster for that family but perhaps for the whole challenge community an even bigger disaster and it starts to feel like [the unethical HCS] around the time of the Second World War. [Scientist, UK/Europe]

3.5.4.1   Generalisability of Adult Challenge Studies to Children

If children are excluded from HCS aimed at understanding and preventing infections that predominantly cause harm to children, then, for pathogens that mainly affect this population, researchers should arguably select participants such that the findings (e.g., regarding vaccine efficacy in the target population) can be generalised to children in endemic settings as reliably as possible (even if there remain important differences between children and adults with respect to the infection in question and/or vaccine efficacy estimates). This raises a further difficulty: children in endemic settings share certain characteristics with adults in similar endemic settings (e.g., genetics, microbiome, etc.); but, with respect to immunity, they may actually be more similar to adults in non-endemic settings (since both young children in endemic settings and most adults in non-endemic settings will be non-immune to the pathogen in question).

Because many adults in endemic settings will have (partial) immunity to the pathogen in question, it may be difficult to recruit non-immune adults locally. Thus it may not always be clear whether it would be (ethically and scientifically) preferable to conduct a given HCS design with adult volunteers from a LMIC in which the pathogen is endemic or in those from a non-endemic (LMIC or HIC) population. An ideal approach, with respect to the scientific aim of generalisability, would be to recruit non-immune (ideally never exposed) adults from a population in a non-endemic area of a LMIC (rather than, for example, from a geographically distant and genetically different HIC population) that would also be closer to a population of children in an endemic area in other respects (e.g., genetics, microbiome). From a pragmatic point of view, it may sometimes be difficult to identify and/or recruit a sufficient number of such individuals (especially if non-immune adults are a rarity in areas near endemic settings), in which case HIC HCS might be more justifiable (see Box 3.25).

Box 3.25 Alternatives to HCS in children

[D]oing something in children with a challenge model, when you could do that in adults, or in naive individuals in [a] developed country, to me, doesn't feel justifiable. [Scientist, UK/Europe]

[S]cientifically, the advantage of working in a developing country is where the relevance of the challenge model is increased by working in the relevant population; and, so, in the context of a disease which primarily affects children, studying adults in an endemic setting, in a developing country, may not be a very good model for understanding the disease in children and it could be that studying healthy volunteers in Australia or in the UK, who are naive to that disease, may be more relevant to children. [Scientist, UK/Europe]

I think [enrolling children in HCS] is really quite likely to have a major adverse effect on the public perception of research … even if you could find parents that would consent … I would be very concerned about that. I think, this is the one area [for] which [there] might be some justification for doing the challenge studies in non-endemic countries … [T]he most practical way of providing data that is relevant to young children, at least in the case of malaria, is with non-endemic adults, because in both cases they're non-immune. [Scientist, UK/Europe]

3.5.4.2   Recent Example of a Low Risk Challenge Study in Children

One recent study of a rotavirus vaccine in South Africa used an HCS design, although the authors do not refer to it as a challenge study (Groome et al. 2017). The study recruited healthy children aged 2–3 years with the aim of testing the safety and immunogenicity of an inactivated injectable rotavirus vaccine; the injectable vaccine contained rotavirus proteins (in contrast to live-attenuated rotavirus vaccines that contain a complete virus potentially capable of replication). After vaccination (or placebo), the children were challenged with a live-attenuated rotavirus vaccine strain, and investigators tested (among other outcomes) whether the inactivated injectable vaccine reduced the shedding (in stool) of the live-attenuated rotavirus vaccine strain. There was some evidence that the inactivated vaccine reduced such shedding (thus, if such an effect were generalisable from the live-attenuated rotavirus strain to wild-type strains, this would suggest that the injectable vaccine might reduce replication of and/or disease resulting from and/or the transmissibility of wild-type rotavirus in those vaccinated and subsequently exposed) (Groome et al. 2017).

It is interesting that the authors did not refer to the above study as a challenge study, although other researchers have identified it as an example of HCS in children (Baay et al. 2018). One might think that it was not considered an HCS design because the challenge agent was not a disease-causing strain of rotavirus (i.e., it was attenuated to a degree that, like other live-attenuated vaccines, it would be expected to lead to immunity to the infection in question without itself causing symptoms/disease). However, HCS need not always be designed to result in symptoms/disease in volunteers (See, for example, Sect. 3.3.2.1), and HCS using (asymptomatic) infection/viraemia as the goal/endpoint of challenge (i.e., an 'infection model' involving few or no symptoms among participants as opposed to a 'disease model' involving significant symptoms) have been proposed and/or conducted with arboviruses such as dengue and Zika (Larsen et al. 2015; Durbin and Whitehead 2017).

In any case, the rotavirus study discussed above does not set a precedent for disease-causing HCS in children since the challenge agent used (a live-attenuated vaccine strain) would (i) entail minimal, if any, risk to participants, and (ii) potentially provide direct benefits to participants if the live-attenuated vaccine strain provided/increased immunity against wild-type infection.

HCS with such highly-attenuated strains (whether designed as vaccines or not) might not be especially generalisable to wild-type infection, which would in some cases undermine the scientific rationale for the use of such designs (see Sect. 3.2.1.1) (Selgelid and Jamrozik 2018). However, if it were shown that an attenuated strain that is incapable of causing disease among participants nevertheless provided useful generalisable knowledge regarding infection with wild-type strains, such a strain could potentially be ideal for use in HCS (in children or in adults) with respect to the assessment of the relative burdens and benefits of such a design.

3.6 Payment of Participants

Enrolling in a challenge study often entails (i) potential financial costs for participants (e.g., travel costs, childcare, and time away from usual activities, including paid work), (ii) potential burdens during participation (including, for example, exposure to risk and/or harm and potentially long periods of isolation—see Sect. 3.3). Since research participants endure these costs and burdens contributing to projects that primarily aim to benefit others (e.g., future people at risk of the disease being studied), it has been argued that payment is often ethically appropriate—although payment of research participants remains controversial, especially in the contexts involving (economically) vulnerable populations and/or high levels of payment (Macklin 1981; McNeill 1997; Savulescu 2001; Cryder et al. 2010; Gelinas et al. 2018). Importantly, payment may be intended as (i) reimbursement for costs incurred, (ii) compensation for harms (if they occur), (iii) compensation for other burdens, (iv) an incentive for participation, or some combination of all of these goals (Gelinas et al. 2018).

Attitudes toward the payment of research participants vary across individuals and across different cultures/countries, as do payment practices. Small reimbursements (e.g., for travel costs) are relatively uncontroversial and quite common across jurisdictions. The need for compensation for research-related harms (if they occur) is also relatively uncontroversial, and was widely supported among interviewees for this project, although legislation and current practices regarding such payments vary considerably in different countries (Chingarande and Moodley 2018). For example, LMIC researchers may sometimes have difficulty obtaining the necessary insurance to cover such compensation for harm, which can be one factor that undermines local research capacity (see Colombian vivax HCS reviewed in Sect. 5.3).

Payment intended to compensate for other burdens and/or incentivise research participation, however, are more controversial (Gelinas et al. 2018). Such payments, and even high levels of payment, are widely viewed as appropriate in some countries (e.g., UK and USA (Savulescu 2001; Cryder et al. 2010)) but they are less accepted in other countries and are sometimes even proscribed by local regulations and/or norms (e.g., in some Latin American countries). Recent HCS have sometimes involved high levels of payment; for example, an HIC influenza HCS offered participants USD $4,000 (Cohen 2016). Among HCS in LMICs reviewed below, participants were paid (i) in Kenya, up to approximately USD $500 (depending on duration of infection and other factors) (Hodgson et al. 2015; Nordling 2018), (ii) in Thailand, amounts indexed to local wages (Thai minimum wage is equal to approximately USD $10 per day,[14] meaning that payment for participation in a one month inpatient study might amount to approximately USD $300 or more), and (iii) in Colombia, no payment apart from reimbursement for costs (travel etc.). Stakeholders interviewed for this report generally agreed that payment for participation in (LMIC) HCS was ethically acceptable, and that it was particularly appropriate to pay individuals who participate in particularly burdensome HCS designs (e.g., inpatient studies). However, determination of the appropriate method(s) for titration of payment according to the burdens of participation was an area that was considered contentious and/or unresolved (see Box 3.26).

Arguments in favour of payment of participants have focused on (i) reciprocity (because participants take on burdens and risks in order to benefit others) (Njue et al. 2018), (ii) analogies between research participation and other forms of labour (Gelinas et al. 2018), (iii), payment for taking on risk as for other types of high-risk socially beneficial activities (e.g., fire-fighting) (Savulescu 2001), (iv) evidence that payment may be a signal that reminds potential participants that studies in healthy volunteers are usually not beneficial for them and may impose a net risk of harm (Cryder et al. 2010), (v) a potentially important incentive to increase participant enrolment numbers (in socially valuable, appropriately low-risk research) (Macklin 1981). As mentioned above, it has also been argued that there should be a system to compensate those who suffer (rare but potentially significant) harms as a result of HCS participation, even if risks have been minimised and fully disclosed to participants (Bambery et al. 2015).

Box 3.26 payment in LMICs

[T]he current situation is … for whatever reason, [that] somebody says it's not appropriate to pay people in developing countries … and sometimes that didn't come from developing countries themselves, it comes from somewhere else and it's

---

[14]Current Thai minimum wage was 325 THB at time of writing, see https://tradingeconomics.com/thailand/minimum-wages [Accessed 30 March 2019].

also perfectly clear that the amounts that are paid in some US settings are grossly inappropriate. [Scientist, UK/Europe]

[I]n endemic countries I think ... when you're asking people to give up ten days of their life, to stay in an inpatient unit ... I think they should be compensated. That's ... a lot of their time and freedom. Certainly it's time away from how they could be making other money – and it's difficult, frankly, to find people altruistic enough [to] say 'Sure, I'll stay in your inpatient unit for ten days, for the betterment of science.' Prof. Anna Durbin, scientist, USA

[F]or a long time, in a [low-income] setting [the standard view has been that] people should not be compensated, so that they can make a voluntary decision not driven by gains that might accrue from participating in the study ... [A]s much as people get worried about [payment in LMICs], it is the same as what you are seeing with people who are doing the phase one studies in Europe ... [T]he students end up doing that [i.e., serving as participants], because they want some extra money [and] because they want to be a part of something. [Scientist, Africa]

[We]'re thinking [that] because we need a population that ... has a better chance of understanding the study, it would be people around [the research institution] – so either hospital staff, students, lecturers, so people who are in the campus probably ... so they won't go for minimum wage, it's crazy right? Ah, so that's compensation for burdens and then there should be incentives. I mean this is a challenge study, no incentives, forget it. No-one's going to come and I think we should incentivise people properly. [Scientist, Asia]

[H]ow do we appropriately compensate for all these inconveniences? And I think tied to that is, and this is what we struggle with, is are we compensating for the risk? Are we compensating for the fact that the inmate actually gets sick and feels all the discomfort that comes with the sickness? How do you even know what to compensate for that? You can compensate for time stayed, you can compensate for expenses, you can fund expenses, but how do you compensate for someone being sick? [Scientist, Africa]

I don't want to underpay people because it's not fair to underpay people. If I take two weeks out [to participate in research], committing this amount of time and sacrificing my social life, [and] probably [drinking] no alcohol for two weeks, [that would constitute a significant burden]. Come on, you have to pay people! [Scientist, Asia]

Often the procedures for challenge studies are really quite onerous compared to other studies so if you just add all that up together, just logically, the amount that they should be paid is more than for other studies. How much that should be, should probably be linked to local purchasing parity. That makes sense to me. [Scientist, UK/Europe]

3.6.1 Undue Inducement

Many researchers and ethicists might be concerned that payment of participants (particularly high levels of payment, and/or among economically disadvantaged populations) may result in 'undue inducement' to enrol in research. This presupposes that some payments or inducements may be appropriate or 'due' (for example, reimbursements of travel costs or payments *in lieu* of lost wages) whereas others are inappropriate (Macklin 1981). Undue financial inducement might (or might be perceived to) (i) undermine the understanding and/or voluntariness components of informed consent by impairing decision-making (leading, for example, to participants accepting more risk than they would usually accept and/or forsaking responsibilities to children and family members) (Grady 2001; Njue et al. 2014, 2018), and/or (ii) lead participants to conceal important details of their medical or psychiatric history resulting in increased risks to them and/or perhaps compromising the scientific validity of the study (Gelinas et al. 2018; Taylor and Morales 2018), and/or (iii) lead participants to 'over-volunteer', e,g., participate in multiple studies simultaneously, which could have similar deleterious effects (see Box 3.27). In contrast, some have argued that—because each of these ethical concerns can be appropriately remedied without removing financial payment—payment per se is not ethically problematic, and even very high levels of payment may be acceptable (Savulescu 2001; Emanuel 2005).

Many of our interviewees were concerned about the potential for undue inducement to participate in HCS (in both HICs and LMICs) and acknowledged that higher levels of poverty, as well as cultural norms, could alter local perceptions of payment for research participation. However, some also argued that this should not necessarily preclude the payment of participants in LMICs, especially for highly burdensome (and/or highly socially valuable) research including some HCS designs. Several noted the potential for undue inducement among economically vulnerable participants in HICs, emphasising that this potential issue was not unique to LMICs, but one that required further work in different settings to determine and review fair levels of payment (see Box 3.27).

Box 3.27 Undue inducement and vulnerable populations

[O]ur ethics committees are very worried about compensation and inducement … But if you look at challenge studies in the West, you look at challenge studies in Africa, whoever is [participating in] the study always says, 'My main motivation is the money.' But, has anybody set a price on what is enough, and what is insufficient? … [I]f I paid someone $100 a day in Baltimore would they participate? If I paid them $1000 a day in Baltimore would they participate? Where do you set the price? … In India or in Africa or anywhere else, I think you should compensate people who are volunteering, and I think you should compensate them well. Inpatient studies should definitely be compensated more than outpatient studies but … I don't think we've done enough work on what's right, and what is actually inducement. [Gagandeep Kang, scientist, India]

[Concerns about undue inducement of vulnerable participants] very rarely came up … in [a research site conducting HCS in UK/Europe]. We mainly had excessively educated, sort of, philosophy students and that kind of thing who found it very interesting, and I think [to] a lot of them … the amount of money we paid really didn't make much difference to them … whereas in [a research site conducting HCS in North America], I've been there and witnessed how they do things there … [and]

it's some of the most vulnerable people in that society [who end up participating]. [Scientist, UK/Europe]

I don't have anything against what they are doing in the States. [The] situation might be completely different; [the] country is completely different. So I don't, I don't think the payment is harming anybody; and … from the volunteers that I have seen in the States, I think that they accept the money because they like the money, but not because they need it … It's not that you are influencing them by—possibly, I mean, who knows? It's very difficult to know. [Sócrates Herrera, scientist, Colombia]

[T]he one worry I have is that if you pay people a lot of money that you could have a higher increase in people not revealing certain information to the study team and lying about inclusion criteria, exclusion criteria, or what they did in the course of the trial, or what side effects they're experiencing. And all of those things could have a detrimental impact on safety. So I think we probably need more data on whether higher payments induce people to conceal information that would relate to their own protection. [Ethicist, North America]

3.6.2 Other Ethical Issues Related to Payment

In addition to undue inducement, there may sometimes be other ethically concerning effects of (high levels of) payment. Firstly, payment may lead to the disproportionate recruitment of impoverished individuals and groups, which would in some cases arguably be unjust, for example if the disease under study did not primarily occur in similar populations (see 'Participant selection') (Macklin 1981; Elliott and Abadie 2008). Secondly, high payment may undermine sustainable research practices and fair availability of research opportunities, an effect to which institutions in resource-limited settings may be particularly susceptible; for example, if certain types of research involve much greater payments, it may be more difficult for other studies (with lower levels of payment) to enrol participants; in the longer term, some worry that this could jeopardise participation in research more generally (Njue et al. 2014). Finally, some worry that payment may change the way that researchers treat research participants and/or that participants will view themselves as being akin to employees as opposed to volunteers (Njue et al. 2014) (see Box 3.28).

There are already fears of creating an 'underclass' of research participants in HICs (drawn from underprivileged groups in society) (Elliott and Abadie 2008) and/or of 'over-volunteering' (i.e. participating in research too frequently in order to receive payment, see Sect. 4.3.4). These patterns of participant selection could in some cases undermine the safety of participants (e.g., because compounds used in one study interact with those used in another study) and/or scientific analyses (e.g., because data generated with the research 'underclass' are not generalisable to the eventual target population) (Shamoo and Resnik 2006; Allen et al. 2017). Such concerns could be magnified in LMIC settings and/or impoverished communities in HICs, especially where there are large populations of unemployed individuals and a fragile social support system. Further social science research on potential inducement of participants and/or over-volunteering as well as transparent longitudinal data (e.g., regarding the level of payment in different studies and the sustainability of research with variable levels of payment over time) would help to clarify the practical importance of such concerns (see Box 3.28).

Payment in both HIC and LMICs is a significant motivator for participation, often rated by participants as more important than the altruistic motives of contributing to important science (though such motivations commonly co-exist) (Njue et al. 2018; Kraft et al. 2019). In endemic settings, where many members of the local population may be more likely to be poor (and/or vulnerable in other respects), even carefully considered levels of payment have caused controversy. For example, in June 2018 a Kenyan media article expressed concern regarding payment of participants in a malaria HCS, despite prior community engagement and thorough consideration of appropriate levels of compensation of appropriate levels of compensation (Gathura 2018; Kenya Medical Research Institute (KEMRI) 2018). In the interviews for this project, payment of HCS participants was identified as a complex issue in need of further analysis (see Box 3.28).

Box 3.28 Sustainability, over-volunteering, and relationships between researchers and participants

There are institutions that are involved, and look at what should be acceptable … [W]ithin our setting, what can we possibly keep up with, and what can we sustain? … [O]nce you start compensating people at a … certain level, they will expect that to continue. And when it comes to another study they will not [have the same levels of compensation], so I think it's one way [of] looking at what would be acceptable within our frameworks, I think, if you are driven by local institutions [and their views on sustainable payment levels]. [Scientist, Africa]

[T]here are several … phase one trials that are happening in India where we have these professional trial participants. They make a livelihood out of trial participation … [T]here is a washout period of 45 days or something and … after every … 45 days, they just go and participate in one trial after another; and these people, if there is no trial, if they are not eligible to participate in a trial, they go hungry. [Vijayaprasad Gopichandran, ethicist, India]

[Overvolunteering] will undermine the science, but I think the primary thing is … thinking from a more society than science perspective … [w]hat it winds up doing is giving all of research a bad name. So the fact that your own research got ruined is bad enough, but you are ruining then research for a number of different areas. [Gagandeep Kang, scientist, India]

[O]ftentimes people think entirely in terms of how paying people might change a person's decision to go into the research, losing sight of the fact that the person that's doing the paying is also altering his or her relationship with that individual by paying them … If you feel like … you can … retain a person in a trial by paying them more, you probably are … feeling much less pressure to treat that person with respect and to try and do various things [to] keep their motivations aligned with your own. [Jonathan Kimmelman, ethicist, Canada]

References

Acosta, P.L., M.T. Caballero, and F.P. Polack. 2016. Brief history and characterization of enhanced respiratory syncytial virus disease. *Clinical and Vaccine Immunology* 23 (3): 189–195.

Allen, C., G. Francis, J. Martin, and M. Boyce. 2017. Regulatory experience of TOPS: An internet-based system to prevent healthy subjects from over-volunteering for UK clinical trials. *European Journal of Clinical Pharmacology* 73 (12): 1551–1555.

Arévalo-Herrera, M., D.A. Forero-Peña, K. Rubiano, J. Gómez-Hincapie, N.L. Martínez, M. Lopez-Perez, A. Castellanos, N. Céspedes, R. Palacios, and J.M. Oñate. 2014. Plasmodium vivax sporozoite challenge in malaria-naive and semi-immune Colombian volunteers. *PLoS ONE* 9 (6): e99754.

Arévalo-Herrera, M., J.M. Vásquez-Jiménez, M. Lopez-Perez, A.F. Vallejo, A.B. Amado-Garavito, N. Céspedes, A. Castellanos, K. Molina, J. Trejos, and J. Oñate. 2016. Protective efficacy of Plasmodium vivax radiation-attenuated sporozoites in Colombian volunteers: A randomized controlled trial. *PLoS Neglected Tropical Diseases* 10 (10): e0005070.

Baay, M.F.D., T.L. Richie, P. Neels, M. Cavaleri, R. Chilengi, D. Diemert, S.L. Hoffman, R. Johnson, B.D. Kirkpatrick, and I. Knezevic. 2018. Human challenge trials in vaccine development, Rockville, MD, USA, September 28–30, 2017. *Biologicals*.

Bambery, B., M. Selgelid, C. Weijer, J. Savulescu, and A.J. Pollard. 2015. Ethical criteria for human challenge studies in infectious diseases. *Public Health Ethics* 9 (1): 92–103.

Battin, M.P., L.P. Francis, J.A. Jacobson, and C.B. Smith. 2008. The ethics of research in infectious disease: Experimenting on this patient, risking harm to that one. In *The patient as victim and vector: Ethics and infectious disease*. Oxford University Press.

Bennett, J.W., B.S. Pybus, A. Yadava, D. Tosh, J.C. Sousa, W.F. McCarthy, G. Deye, V. Melendez, and C.F. Ockenhouse. 2013. Primaquine failure and cytochrome P-450 2D6 in Plasmodium vivax malaria. *New England Journal of Medicine* 369 (14): 1381–1382.

Bonham, V.H., and J.D. Moreno. 2008. Research with captive populations: Prisoners, students, and soldiers.

Chattopadhyay, R., and D. Pratt. 2017. Role of controlled human malaria infection (CHMI) in malaria vaccine development: A US food and drug administration (FDA) perspective. *Vaccine* 35 (21): 2767.

Chingarande, G.R., and K. Moodley. 2018. Disparate compensation policies for research related injury in an era of multinational trials: A case study of Brazil, Russia, India, China and South Africa. *BMC Medical Ethics* 19 (1): 8.

Cohen, J. 2016. Studies that intentionally infect people with disease-causing bugs are on the rise. In *Science magazine*. Washington DC, USA, American Academy for the Advancement of Science.

Collins, K.A., C.Y.T. Wang, M. Adams, H. Mitchell, M. Rampton, S. Elliott, I.J. Reuling, T. Bousema, R. Sauerwein, and S. Chalon. 2018. A controlled human malaria infection model enabling evaluation of transmission-blocking interventions. *The Journal of Clinical Investigation*.

Cryder, C.E., A.J. London, K.G. Volpp, and G. Loewenstein. 2010. Informative inducement: Study payment as a signal of risk. *Social Science and Medicine* 70 (3): 455–464.

Darton, T.C., C.J. Blohmke, V.S. Moorthy, D.M. Altmann, F.G. Hayden, E.A. Clutterbuck, M.M. Levine, A.V.S. Hill, and A.J. Pollard. 2015. Design, recruitment, and microbiological considerations in human challenge studies. *The Lancet Infectious Diseases* 15 (7): 840–851.

Durbin, A.P., and S.S. Whitehead. 2017. Zika vaccines: Role for controlled human infection. *The Journal of Infectious Diseases* 216 (Suppl 10): S971–S975.

Edwards, S.J.L. 2005. Research participation and the right to withdraw. *Bioethics* 19 (2): 112–130.

El Setouhy, M., T. Agbenyega, F. Anto, C.A. Clerk, K.A. Koram, M. English, R. Juma, C. Molyneux, N. Peshu, and N. Kumwenda. 2004. Moral standards for research in developing countries from "reasonable availability" to "fair benefits". *The Hastings Center Report* 34 (3): 17–27.

Elliott, C., and R. Abadie. 2008. Exploiting a research underclass in phase 1 clinical trials. *New England Journal of Medicine* 358 (22): 2316–2317.

Emanuel, E.J. 2005. Undue inducement: Nonsense on stilts? *The American Journal of Bioethics* 5 (5): 9–13.

Emanuel, E.J., G. Bedarida, K. Macci, N.B. Gabler, A. Rid, and D. Wendler. 2015. Quantifying the risks of non-oncology phase I research in healthy volunteers: Meta-analysis of phase I studies. *BMJ* 350: h3271.

Evers, D.L., C.B. Fowler, J.T. Mason, and R.K. Mimnall. 2015. Deliberate microbial infection research reveals limitations to current safety protections of healthy human subjects. *Science and Engineering Ethics* 21 (4): 1049–1064.

Eyal, N., M. Lipsitch, T. Bärnighausen, and D. Wikler. 2018. Opinion: Risk to study nonparticipants: A procedural approach. *Proceedings of the National Academy of Sciences* 115 (32): 8051–8053.

Gathura, G. 2018. Want cash? Volunteer for a dose of malaria parasite, says Kemri amid ethical queries. *The Standard*. Kenya, Standard Group PLC.

Gelinas, L., E.A. Largent, I.G. Cohen, S. Kornetsky, B.E. Bierer, and H. Fernandez Lynch. 2018. A framework for ethical payment to research participants. *New England Journal of Medicine* 378 (8).

Gibani, M.M., C. Jin, T.C. Darton, and A.J. Pollard. 2015. Control of invasive Salmonella disease in Africa: Is there a role for human challenge models? *Clinical Infectious Diseases* 61 (Suppl 4): S266–S271.

Goodin, R.E. 1986. *Protecting the vulnerable: A re-analysis of our social responsibilities*. University of Chicago Press.

Goodyear, M. 2006. Learning from the TGN1412 trial. *British Medical Journal Publishing Group*.

Gordon, S.B., J. Rylance, A. Luck, K. Jambo, D.M. Ferreira, L. Manda-Taylor, P. Bejon, B. Ngwira, K. Littler, and Z. Seager. 2017. A framework for controlled human infection model (CHIM) studies in Malawi: Report of a Wellcome Trust workshop on CHIM in low income countries held in Blantyre, Malawi. *Wellcome Open Research* 2.

Grady, C. 2001. Money for research participation: Does it jeopardize informed consent? *American Journal of Bioethics* 1 (2): 40–44.

Groome, M.J., A. Koen, A. Fix, N. Page, L. Jose, S.A. Madhi, M. McNeal, L. Dally, I. Cho, and M. Power. 2017. Safety and immunogenicity of a parenteral P2-VP8-P [8] subunit rotavirus vaccine in toddlers and infants in South Africa: A randomised, double-blind, placebo-controlled trial. *The Lancet Infectious Diseases* 17 (8): 843–853.

Helgesson, G., and L. Johnsson. 2005. The right to withdraw consent to research on biobank samples. *Medicine, Health Care and Philosophy* 8 (3): 315–321.

Herrera, S., O. Fernández, M.R. Manzano, B. Murrain, J. Vergara, P. Blanco, R. Palacios, J.D. Vélez, J.E. Epstein, and M. Chen-Mok. 2009. Successful sporozoite challenge model in human volunteers with Plasmodium vivax strain derived from human donors. *The American Journal of Tropical Medicine and Hygiene* 81 (5): 740–746.

Herrera, S., Y. Solarte, A. Jordán-Villegas, J.F. Echavarría, L. Rocha, R. Palacios, Ó. Ramírez, J.D. Vélez, J.E. Epstein, and T.L. Richie. 2011. Consistent safety and infectivity in sporozoite challenge model of Plasmodium vivax in malaria-naive human volunteers. *The American Journal of Tropical Medicine and Hygiene* 84 (Suppl 2): 4–11.

Herrington, D.A., L. Van De Verg, S.B. Formal, T.L. Hale, B.D. Tall, S.J. Cryz, E.C. Tramont, and M.M. Levine. 1990. Studies in volunteers to evaluate candidate Shigella vaccines: Further experience with a bivalent Salmonella typhi-Shigella sonnei vaccine and protection conferred by previous Shigella sonnei disease. *Vaccine* 8 (4): 353–357.

Hodgson, S.H., E. Juma, A. Salim, C. Magiri, D. Kimani, D. Njenga, A. Muia, A.O. Cole, C. Ogwang, and K. Awuondo. 2014. Evaluating controlled human malaria infection in Kenyan adults with varying degrees of prior exposure to Plasmodium falciparum using sporozoites administered by intramuscular injection. *Frontiers in Microbiology* 5: 686.

Hodgson, S.H., E. Juma, A. Salim, C. Magiri, D. Njenga, S. Molyneux, P. Njuguna, K. Awuondo, B. Lowe, and P.F. Billingsley. 2015. Lessons learnt from the first controlled human malaria infection study conducted in Nairobi, Kenya. *Malaria Journal* 14 (1): 182.

Hope, T., and J. McMillan. 2004. Challenge studies of human volunteers: Ethical issues. *Journal of Medical Ethics* 30 (1): 110–116.

Jin, C., M.M. Gibani, M. Moore, H.B. Juel, E. Jones, J. Meiring, V. Harris, J. Gardner, A. Nebykova, and S.A. Kerridge. 2017. Efficacy and immunogenicity of a Vi-tetanus toxoid conjugate vaccine in the prevention of typhoid fever using a controlled human infection model of Salmonella Typhi: A randomised controlled, phase 2b trial. *The Lancet*.

Johnson, R.A., A. Rid, E. Emanuel, and D. Wendler. 2016. Risks of phase I research with healthy participants: A systematic review. *Clinical Trials* 13 (2): 149–160.

Jongo, S.A., S.A. Shekalaghe, L.W.P. Church, A.J. Ruben, T. Schindler, I. Zenklusen, T. Rutishauser, J. Rothen, A. Tumbo, and C. Mkindi. 2018. Safety, immunogenicity, and protective efficacy against controlled human malaria infection of Plasmodium falciparum sporozoite vaccine in Tanzanian adults. *The American Journal of Tropical Medicine and Hygiene* 99 (2): 338–349.

Kamm, F.M. 1989. Harming some to save others. *Philosophical Studies* 57 (3): 227–260.

Kenter, M.J.H., and A.F. Cohen. 2006. Establishing risk of human experimentation with drugs: Lessons from TGN1412. *The Lancet* 368 (9544): 1387–1391.

Kenya Medical Research Institute (KEMRI). 2018. Response to an article carried in The Standard. Nairobi, Kenya, KEMRI.

Kimmelman, J. 2005. Medical research, risk, and bystanders. *IRB* 27 (4): 1.

Kiwanuka, O., B.-M. Bellander, and A. Hånell. 2018. The case for introducing pre-registered confirmatory pharmacological pre-clinical studies. *Journal of Cerebral Blood Flow and Metabolism* 38 (5): 749–754.

Kraft, S.A., D.M. Duenas, J.G. Kublin, K.J. Shipman, S.C. Murphy, and S.K. Shah. 2019. Exploring ethical concerns about human challenge studies: A qualitative study of controlled human malaria infection study participants' motivations and attitudes. *Journal of Empirical Research on Human Research Ethics* 14 (1): 49–60.

Lange, M.M., W. Rogers, and S. Dodds. 2013. Vulnerability in research ethics: A way forward. *Bioethics* 27 (6): 333–340.

Larsen, C.P., S.S. Whitehead, and A.P. Durbin. 2015. Dengue human infection models to advance dengue vaccine development. *Vaccine* 33 (50): 7075–7082.

Lell, B., B. Mordmüller, J.-C.D. Agobe, J. Honkpehedji, J. Zinsou, J.B. Mengue, M.M. Loembe, A.A. Adegnika, J. Held, and A. Lalremruata. 2017. Impact of sickle cell trait and naturally acquired immunity on uncomplicated malaria after controlled human malaria infection in adults in Gabon.

Levine, M., R. Black, C. Ferreccio, R. Germanier, and C.T. Committee. 1987. Large-scale field trial of Ty21a live oral typhoid vaccine in enteric-coated capsule formulation. *The Lancet* 329 (8541): 1049–1052.

London, A.J. 2005. Undue inducements and reasonable risks: Will the dismal science lead to dismal research ethics? *The American Journal of Bioethics* 5 (5): 29–32.

London, A.J. 2006. Reasonable risks in clinical research: A critique and a proposal for the integrative approach. *Statistics in Medicine* 25 (17): 2869–2885.

London, A.J. 2007. Two dogmas of research ethics and the integrative approach to human-subjects research. *The Journal of Medicine and Philosophy* 32 (2): 99–116.

London, A.J., and J. Kimmelman. 2019. Clinical trial portfolios: A critical oversight in human research ethics, drug regulation, and policy. *Hastings Center Report* 49 (4): 31–41.

Luna, F. 1997. Vulnerable populations and morally tainted experiments. *Bioethics* 11 (3–4): 256–264.

Luna, F. 2009. Elucidating the concept of vulnerability: Layers not labels. *IJFAB: International Journal of Feminist Approaches to Bioethics* 2 (1): 121–139.

Lynch, H.F. 2012. The rights and wrongs of intentional exposure research: Contextualising the Guatemala STD inoculation study. *Journal of Medical Ethics* 38 (8): 513–515.

Macklin, R. 1981. 'Due' and 'Undue' inducements: On paying money to research subjects. *IRB: Ethics and Human Research* 3 (5): 1–6.

Macklin, R. 2003. Bioethics, vulnerability, and protection. *Bioethics* 17 (5–6): 472–486.

Malaria Vaccine Initiative. 2016. The challenges of malaria vaccine "challenge" trials: Mosquitoes travel in business class to infect American volunteers—but fare better in economy. https://www.malariavaccine.org/news-events/news/challenges-malaria-vaccine-challenge-trials-mosquitoes-travel-business-class. Accessed 14 Feb 2019.

Mammen, M.P., A. Lyons, B.L. Innis, W. Sun, D. McKinney, R.C.Y. Chung, K.H. Eckels, R. Putnak, N. Kanesa-Thasan, and J.M. Scherer. 2014. Evaluation of dengue virus strains for human challenge studies. *Vaccine* 32 (13): 1488–1494.

McConnell, T. 2010. The inalienable right to withdraw from research. *The Journal of Law, Medicine and Ethics* 38 (4): 840–846.

McCullagh, D., H.C. Dobinson, T. Darton, D. Campbell, C. Jones, M. Snape, Z. Stevens, E. Plested, M. Voysey, and S. Kerridge. 2015. Understanding paratyphoid infection: Study protocol for the development of a human model of Salmonella enterica serovar Paratyphi A challenge in healthy adult volunteers. *British Medical Journal Open* 5 (6): e007481.

McNeill, P. 1997. Paying people to participate in research: Why not? *Bioethics* 11 (5): 390–396.

Meltzer, L.A., and J.F. Childress. 2008. What is fair participant selection? In *The Oxford textbook of clinical research ethics*, ed. Ezekiel J. Emanuel, Christine Grady, Robert A. Crouch, Reidar K. Lie, Franklin G. Miller, and David Wendler, 377–385. Oxford: Oxford University Press.

Miller, F.G. 2003. Ethical issues in research with healthy volunteers: Risk-benefit assessment. *Clinical Pharmacology and Therapeutics* 74 (6): 513–515.

Miller, F.G., and C. Grady. 2001. The ethical challenge of infection-inducing challenge experiments. *Clinical Infectious Diseases* 33 (7): 1028–1033.

Miller, F.G., and S. Joffe. 2009. Limits to research risks. *Journal of Medical Ethics* 35 (7): 445–449.

Miller, F.G., and D.L. Rosenstein. 2008. Challenge experiments. In: *The Oxford textbook of clinical research ethics*, 273–279.

Moore, N. 2016. Lessons from the fatal French study BIA-10-2474. *BMJ: British Medical Journal (Online)* 353.

National Commission for the Proptection of Human Subjects of Biomedical and Behavioral Research, B.M. 1978. *The Belmont report: Ethical principles and guidelines for the protection of human subjects of research*. Superintendent of Documents.

Nieman, A.-E., Q. de Mast, M. Roestenberg, J. Wiersma, G. Pop, A. Stalenhoef, P. Druilhe, R. Sauerwein, and A. van der Ven. 2009. Cardiac complication after experimental human malaria infection: A case report. *Malaria Journal* 8 (1): 277.

Njue, M., F. Kombe, S. Mwalukore, S. Molyneux, and V. Marsh. 2014. What are fair study benefits in international health research? Consulting community members in Kenya. *PLoS ONE* 9 (12): e113112.

Njue, M., P. Njuguna, M.C. Kapulu, G. Sanga, P. Bejon, V. Marsh, S. Molyneux, and D. Kamuya. 2018. Ethical considerations in controlled human malaria infection studies in low resource settings: Experiences and perceptions of study participants in a malaria challenge study in Kenya. *Wellcome Open Research* 3.

Nordling, L. 2018. The ethical quandary of human infection studies. https://undark.org/article/ethical-quandry-human-infection/#comments. Accessed 16 Mar 2019.

Olotu, A., V. Urbano, A. Hamad, M. Eka, M. Chemba, E. Nyakarungu, J. Raso, E. Eburi, D.O. Mandumbi, and D. Hergott. 2018. Advancing global health through development and clinical trials partnerships: A randomized, placebo-controlled, double-blind assessment of safety, tolerability, and immunogenicity of PfSPZ vaccine for malaria in healthy equatoguinean men. *The American Journal of Tropical Medicine and Hygiene* 98 (1): 308–318.

Orjuela-Sanchez, P., Z.H. Villa, M. Moreno, C. Tong-Rios, S. Meister, G.M. LaMonte, B. Campo, J.M. Vinetz, and E.A. Winzeler. 2018. Developing plasmodium vivax resources for liver stage study in the Peruvian Amazon region. *ACS Infectious Diseases* 4 (4): 531–540.

Paul, Y. 2004. Herd immunity and herd protection. *Vaccine* 22 (3): 301–302.

Pitisuttithum, P. 2018. Controlled human infection model (workshop presentation). In *Towards a new ethical framework for the use of human challenge studies on emerging infectious diseases*. Brocher Foundation.

Pollard, A.J., J. Savulescu, J. Oxford, A.V.S. Hill, M.M. Levine, D.J.M. Lewis, R.C. Read, D.Y. Graham, W. Sun, and P. Openshaw. 2012. Human microbial challenge: The ultimate animal model. *The Lancet Infectious Diseases* 12 (12): 903–905.

Pratt, B., D. Zion, K.M. Lwin, P.Y. Cheah, F. Nosten, and B. Loff. 2012. Closing the translation gap for justice requirements in international research. *Journal of Medical Ethics* 38 (9): 552–558.

Resnik, D.B. 2005. Eliminating the daily life risks standard from the definition of minimal risk. *Journal of Medical Ethics* 31 (1): 35–38.

Rid, A. 2014. Setting risk thresholds in biomedical research: Lessons from the debate about minimal risk. *Monash Bioethics Review* 32 (1–2): 63–85.

Rid, A., E.J. Emanuel, and D. Wendler. 2010. Evaluating the risks of clinical research. *JAMA* 304 (13): 1472–1479.

Robinson, W.M., and B.T. Unruh. 2008. The hepatitis experiments at the Willowbrook State School. In *The Oxford textbook of clinical research ethics*, 80–85.

Roestenberg, M., M.-A. Hoogerwerf, D.M. Ferreira, B. Mordmüller, and M. Yazdanbakhsh. 2018a. Experimental infection of human volunteers. *The Lancet Infectious Diseases*.

Roestenberg, M., I. Kamerling, and S.J. de Visser. 2018b. Dealing with uncertainty in vaccine development: The malaria case. *Frontiers in Medicine* 5: 297.

Roestenberg, M., I.M.C. Kamerling, and S.J. de Visser. 2018c. Controlled human infections as a tool to reduce uncertainty in clinical vaccine development. *Frontiers in Medicine* 5.

Roestenberg, M., G.A. O'Hara, C.J.A. Duncan, J.E. Epstein, N.J. Edwards, A. Scholzen, A.J.A.M. Van der Ven, C.C. Hermsen, A.V.S. Hill, and R.W. Sauerwein. 2012. Comparison of clinical and parasitological data from controlled human malaria infection trials. *PLoS ONE* 7 (6): e38434.

Rogers, W., C. Mackenzie, and S. Dodds. 2012. Why bioethics needs a concept of vulnerability. *IJFAB: International Journal of Feminist Approaches to Bioethics* 5 (2): 11–38.

Rothman, D.J. 1982. Were Tuskegee and Willowbrook 'studies in nature'? *Hastings Center Report* 5–7.

Saethre, E., and J. Stadler. 2013. Malicious whites, greedy women, and virtuous volunteers: Negotiating social relations through clinical trial narratives in South Africa. *Medical Anthropology Quarterly* 27 (1): 103–120.

Sauerwein, R.W., M. Roestenberg, and V.S. Moorthy. 2011. Experimental human challenge infections can accelerate clinical malaria vaccine development. *Nature Reviews Immunology* 11 (1): 57.

Savulescu, J. 1998. Commentary: Safety of participants in non-therapeutic research must be ensured. *British Medical Journal* 316 (7135): 891–893.

Savulescu, J. 2001. The fiction of "undue inducement": Why researchers should be allowed to pay participants any amount of money for any reasonable research project. *American Journal of Bioethics* 1 (2): 1g–3g.

Schaefer, G.O., and A. Wertheimer. 2010. The right to withdraw from research. *Kennedy Institute of Ethics Journal* 20 (4): 329–352.

Selgelid, M. 2013. The ethics of human microbial challenge (conference paper). In *Controlled human infection studies in the development of vaccines and therapeutics*. Jesus College, Cambridge, UK.

Selgelid, M.J., and E. Jamrozik. 2018. Ethical challenges posed by human infection challenge studies in endemic settings. *Indian Journal of Medical Ethics*.

Shah, S.K., J. Kimmelman, A.D. Lyerly, H.F. Lynch, F. McCutchan, F.G. Miller, R. Palacios, C. Pardo-Villamizar, and C. Zorilla. 2017. Ethical considerations for Zika virus human challenge trials. *National Institute of Allergy and Infectious Diseases*.

Shah, S.K., J. Kimmelman, A.D. Lyerly, H.F. Lynch, F.G. Miller, R. Palacios, C.A. Pardo, and C. Zorrilla. 2018. Bystander risk, social value, and ethics of human research. *Science* 360 (6385): 158–159.

Shamoo, A.E., and D.B. Resnik. 2006. Strategies to minimize risks and exploitation in phase one trials on healthy subjects. *The American Journal of Bioethics* 6 (3): W1–W13.

Shaw, D. 2014. The right to participate in high-risk research. *The Lancet* 383 (9921): 1009–1011.

Sheffield, J.S., R.R. Faden, M.O. Little, A.D. Lyerly, and C.B. Krubiner. 2018. Pregnant women and vaccines against emerging pathogens: Ethics guidance on an inclusive and responsive research agenda and epidemic response. *American Journal of Obstetrics and Gynecology* 219 (6): 650.

Shekalaghe, S., M. Rutaihwa, P.F. Billingsley, M. Chemba, C.A. Daubenberger, E.R. James, M. Mpina, O.A. Juma, T. Schindler, and E. Huber. 2014. Controlled human malaria infection of Tanzanians by intradermal injection of aseptic, purified, cryopreserved Plasmodium falciparum sporozoites. *The American Journal of Tropical Medicine and Hygiene* 91 (3): 471–480.

Taylor, H.A., and C. Morales. 2018. Is it ethically appropriate to refuse to compensate participants who are believed to have intentionally concealed medical conditions? *American Journal of Bioethics* 4: 83–84.

Thomas, S.J. 2013. Dengue human infection model: Re-establishing a tool for understanding dengue immunology and advancing vaccine development. *Human Vaccines and Immunotherapeutics* 9 (7): 1587–1590.

UK Academy of Medical Sciences. 2005. *Microbial challenge studies of human volunteers*. London: Academy of Medical Sciences.

US Department of Health and Human Services. 1979. The Belmont Report.

van Meer, M.P.A., G.J.H. Bastiaens, M. Boulaksil, Q. de Mast, A. Gunasekera, S.L. Hoffman, G. Pop, A.J.A.M. van der Ven, and R.W. Sauerwein. 2014. Idiopathic acute myocarditis during treatment for controlled human malaria infection: A case report. *Malaria Journal* 13 (1): 38.

Wahdan, M.H., C. Serie, R. Germanier, A. Lackany, Y. Cerisier, N. Guerin, S. Sallam, P. Geoffroy, A.S. El Tantawi, and P. Guesry. 1980. A controlled field trial of live oral typhoid vaccine Ty21a. *Bulletin of the World Health Organization* 58 (3): 469.

Wendler, D. 2005. Protecting subjects who cannot give consent: Toward a better standard for "minimal" risks. *Hastings Center Report* 35 (5): 37–43.

Wendler, D., and E.J. Emanuel. 2005. What is a "minor" increase over minimal risk? *The Journal of Pediatrics* 147 (5): 575–578.

Wenner, D.M. 2015. The social value of knowledge and international clinical research. *Developing World Bioethics* 15 (2): 76–84.

Wenner, D.M. 2017. The social value of knowledge and the responsiveness requirement for international research. *Bioethics* 31 (2): 97–104.

WHO Expert Committee on Biological Standardization. 2016. *Human challenge trials for vaccine development: Regulatory considerations*. Geneva: World Health Organization.

Wikler, D. 2017. Must research benefit human subjects if it is to be permissible? *Journal of Medical Ethics* 43 (2): 114–117.

Wilder-Smith, A., J. Hombach, N. Ferguson, M. Selgelid, K. O'Brien, K. Vannice, A. Barrett, E. Ferdinand, S. Flasche, and M. Guzman. 2018. Deliberations of the strategic advisory group of experts on immunization on the use of CYD-TDV dengue vaccine. *The Lancet Infectious Diseases*.

World Health Organization. 2017. *Ethical issues associated with vector-borne diseases: Report of a WHO scoping meeting*. Geneva: WHO.

World Medical Association. 2008. Declaration of Helsinki. Ethical principles for medical research involving human subjects.

Chapter 4
Community Engagement, Ethics Review, and Regulation

4.1 Community Engagement

Community engagement is ethically important for many types of research. Given that HCS represent a particularly complex, sometimes unfamiliar, and potentially controversial type of research, engagement may be especially warranted in the context of HCS (World Health Organization 2017). Since different issues may arise in different communities, and since HCS may be particularly unfamiliar in LMICs, community engagement is arguably an essential part of setting up and maintaining HCS capacity in LMICs (El Setouhy et al. 2004; Njue et al. 2014; Hodgson et al. 2015). The LMIC HCS case studies reviewed below generally occurred within long-established research institutions, some of which had teams specifically appointed to engage with local communities. The group in Kenya has a particularly significant track record of community engagement regarding research in general and, more recently, HCS in particular (Gikonyo et al. 2008; Njue et al. 2014, 2018).

Several interviewees noted that, ideally, engagement does not merely entail researchers informing communities about planned or on-going research, but should be a two-way process from which researchers could also learn about community perspectives, suggestions, or concerns etc. (see Box 4.1). Engagement activities might also involve consultation with other institutional staff, including ethics committee members. Indeed, several stakeholders identified engagement with, and capacity building of, local ethics committees as a key area that had been necessary for the conduct of some HCS (in both HICs and LMICs), especially where those committees had little previous experience of HCS designs (see Box 4.2).

© The Author(s) 2021
E. Jamrozik and M. J. Selgelid, *Human Challenge Studies in Endemic Settings*,
SpringerBriefs in Ethics, https://doi.org/10.1007/978-3-030-41480-1_4

Box 4.1 Community engagement for HCS

[Our trials have a significant] community engagement component ... [O]ne of the first things is to set up a website and ... map up stakeholders ... [W]e have [participants, and] the research ethics guidelines [for them] are quite comprehensive. What about the stakeholders? ... [W]e have ... ethics committees ... and then we have friends and family [of participants], including social media, for instance ... it can go crazy, you know, [and] you need to manage it from the beginning and then you have the media and the wider public ... [A]ll these stakeholders ... they have their interest[s] ... and they would want certain types of information. [Scientist, Asia]

[Regarding community engagement] I'd have more rather than less information. I think that often we have this bad reputation because we create a vacuum of information, a void. And that void gets filled by misunderstanding. So ... I think we ought to be proactive and avoid that void by filling it with information for the community. And openly ... I think that for a number of reasons, not just, you know, pragmatically getting involvement but also for people to understand what is it that comes out of these studies, what the risks are. I mean we say so little about the benefits from research in lower/middle-income countries, in general. [Ethicist, North America]

We are lucky we have ... a group that is called the KEMRI [Kenya Medical Research Institute] Community Representatives. These are people elected. We have two hundred and twenty people that we are interacting with every three months. Just to talk about the work of the KEMRI but also to hear comments from the community ... [T]hose become key people whom community members can go to, ask questions, get clarifications, you know, complain too, if they want to complain, and so they give us this information. [Scientist, Africa]

I think with something like challenge studies where ... there is a potential if messages are half heard, or shared, you know, out of context ... there is a potential [for] rumours, or worries, or, you know, issues to flare up. And so, if you don't have the sort of mechanisms and relationships that allow people to say, "Hold on, I'm actually worried about this or that," or "No, I don't like the fact that this or that is happening." And you're able to discuss it, and do so in a way that isn't defensive or dishonest, I think that's the only way, really, that you can get this kind of complex information discussed, and that there can be mutual learning. So I think it's important for all kinds of studies because I think it's remarkable what kind of activities can lead to concerns that you don't necessarily predict. [Scientist, Africa]

Box 4.2 Engagement with ethics committees and by ethicists

[Wh]at we did in Kenya was to possibly spend a couple of years before we did a challenge study to sensitise the scientific community why we wanted to do challenge studies and [what] we think they [are] really warrant[ed] for, at this point in time, and there was a back and forth. Initially, people were very sceptical; but ... we explained to them more [and] then were able to meet with the ethics committees and explain

> to them about this technology, why we wanted to go through this route, what … it bring[s] to the table for everybody. [Scientist, Africa]
>
> One of our key stakeholder groups here in Thailand, is ethics committees. Not just our own ethics committees, and obviously we have to engage with them properly because they need to look at our protocol and all that. But also making an effort to engage with other ethics committees … because we have a reputational risk, right? The reputational risk for our ethics committee: 'You what? You approve[d] this study?' Other committees might say 'You're crazy!' – so we should engage with them. [Scientist, Asia]
>
> [F]or ethicists in particular, engaging with the community becomes kind of like 'Of course it must be done, yes it must be done', but I don't see them doing it very often. So, it requires a lot of effort; social scientists tend to do a better job than biomedical ethicists. But I think it is something we're going to increasingly need to think about, not just for challenge studies but for all of clinical research. [Gagandeep Kang, scientist, India]

4.2 Ethical Review

HCS are sometimes perceived to be an unusual and/or particularly sensitive type of research, and thus some commentators have recommended policies of special ethics review procedures, for example by a specialised and/or national committee (UK Academy of Medical Sciences 2005; Bambery et al. 2015; Shah et al. 2018). In contrast, some have argued for maintenance of the *status quo*—i.e., that HCS should be reviewed according to normal procedures for research with healthy volunteers—since other kinds of research involving similar levels of risk are adequately handled via ordinary ethics review committee oversight processes (and even though HCS might be particularly complex and/or specialised, committees are generally empowered to appoint experts to assist with review of highly specialised research) (Hope and McMillan 2004) (see Box 4.3).

> **Box 4.3 Standard ethics review of HCS**
>
> [A]t the end of the day, the requirement[s] for a trial just like any other trial are the same … I think the main thing is building the capacity of the ethics committees to know what are the issues around challenge studies, what are the salient issues and what are the emerging issues. [Scientist, Africa]
>
> I think the important thing is that they're reviewed by a committee with sufficient capacity to perform the review, full stop. It doesn't need to be a national committee. But there may be some settings where they don't really have that capacity. They would need to be able to really understand this, and it actually relates to … a general issue about this review, which is how the scientific aspects are reviewed. [Importantly,] somebody needs to be able to look at the scientific basis for the risks and the benefits

... [The committee] needs to have adequate capacity [to review] both the scientific and the ethical ... considerations ... I would say that's the same for all clinical research. [Scientist, UK/Europe]

I would recommend against [specialised ethics review]. I think we would be creating red tape. I think the ... fundamental concept of an IRB applies to any research study ... I don't think there's anything special about a human challenge trial ... any phase I study almost by definition is not going to benefit the participants. [Scientist, North America]

[Y]ou can imagine [that], in science, we'll always have new things coming up ... The issue is how do we capacity build the ethics of your committees to address the new changes that are coming in ... proactively, not wait[ing] for things to happen, for them to catch up with how they review ... I think there's a lot of experience in ethical review, there's a lot of capacity building that has been going on. [Scientist, Africa]

4.2.1 Ethical Frameworks for Human Challenge Studies

One option for enhancing the review of HCS would be the development of specific ethical principles/guidelines/frameworks for HCS (and/or, for example, for HCS with particular pathogen) (Miller and Grady 2001; Selgelid 2013; WHO Expert Committee on Biological Standardization 2016; Davies 2019). Some stakeholders were in favour of such ethical frameworks, whereas others argued that specific considerations related to particular pathogens were more important, and thus that the scientific expertise of ethics committee members (e.g., regarding the pathogen in question) would be more important to ensure the ethical conduct of HCS (see Box 4.4). Among those who favoured the development of specific ethical guidelines/frameworks for HCS, certain ethical issues were identified as potential candidates for inclusion (and/or as issues not covered satisfactorily by existing research ethics frameworks), including (i) limits to risk to participants, (ii) third-party risk management, and (iii) risk-benefit assessments of HCS as compared with alternative study designs (see Box 4.4).

Box 4.4 Ethical frameworks/guidelines for HCS

I think we do need more frameworks and guidelines for human challenge studies because I think they do raise [particularly salient] questions ... and, therefore, require more careful thinking than we've done in the context of other types of trials. [Ethicist, North America]

I wouldn't make it a special framework specific to challenge studies. There are things that arise in challenge studies that might also arise in other contexts that I think call for a different framework. [For example,] bystander risks come up in challenge trials but also come up in other types of research like HIV cure studies. And [bystander risks] are not something that IRBs or the US regulations cover. So you do need a different framework to think about bystander risks. [Ethicist, North America]

[HCS are] a challenge for our [standard research ethics] benefit framework. Is it really the case that, if the question is socially important enough (and [diseases] like malaria and dengue and Zika, are pretty damn socially important), an adult can consent to, in effect, an unlimited amount of [or at least] a very high degree of risk? I think that can't be right. So, I think there has to be a line there somewhere … What that line is to me, is really the hard question of CHIM. [Ethicist, North America]

I think the principles are largely the same as with other types of studies but there are these additional questions that we've actually been debating like [for example] if there's a perfectly good animal model or you've got a very high attack rate in the field and you only need to recruit twenty people in the field, why would you deliberately expose healthy people to the pathogen? So I think there are issues around understanding challenge studies [and] the scientific process which leads into the ethical questions is quite important. [Scientist, UK/Europe]

[Ethics committees reviewing HCS] generally [have] some expertise in the disease that you're working with and [disease-specific expertise] is, I think, more important than having a general framework for all the different challenge models. So I would rather put my protocol in front of a malaria specialist than a generalist in … challenge models … There are different risks for *Shigella*, for *Salmonella*, [and] for malaria – and to generalise those into one framework I think you run the risk of trivialising some of those risks or [by trying to] make a level playing field for everyone you'll [make], say, *Shigella* [on par with] malaria when they're really not on a par. [Scientist, North America]

I think it would probably be useful to have special guidance for low resource settings. I think there are just enough issues around how much infrastructure is enough infrastructure, payment issues, community consultation issues, [etc.]. So I think the idea of guidance that's directed at low resource settings would be useful. [Ethicist, North America]

As discussed earlier (see Chap. 3), the bioethical literature includes examples of ethical frameworks, principles, and/or criteria for HCS (and/or specific issues related to HCS) (Miller and Grady 2001; Bambery et al. 2015; Shah et al. 2018). One framework for HCS review (drawing on such literature and prior experience of ethics review more generally) has been proposed by Hugh Davies as part of the online resource 'Reviewing Research' (Davies 2019). In addition to more general research ethics considerations, Davies highlights the importance of (i) "public consultation, involvement in design and public access to study details" (i.e., community engagement and transparency), given the sensitivity of HCS designs, and (ii) assessment of "harms to possible contacts and the environment" given the risk of transmission of challenge strains (Davies 2019). Given the growing number of HCS in multiple countries, the growing ethics literature on HCS, and the controversial and/or unresolved issues highlighted in this report, there may be a role for ethical frameworks for HCS. Among other related developments, the WHO Global Health Ethics Unit is in the process of developing guidance on ethical issues related to HCS.

4.2.2   Potential Models for Special Ethical Review

Regarding potential special review procedures for HCS, different models have been proposed, including: (i) the appointment of a special committee (e.g., a national committee) or sub-committee for review of all HCS (e.g., with additional infectious disease expertise) (UK Academy of Medical Sciences 2005; Bambery et al. 2015), (ii) special review for *new* challenge models in particular (e.g., with two independent experts, perhaps followed by more usual review for future use of that model, once it is shown to be safe and scientifically valuable) (Bambery et al. 2015; Shah et al. 2017), or (iii) usual review with particularly strict requirements for a prior systematic review, publicly available rationale, and well-defined compensation for harm (all of which might be required for other kinds of studies, but could perhaps be more strictly required in the case of HCS) (Bambery et al. 2015).

Ultimately, policymakers in any given jurisdiction will need to adopt a policy regarding HCS review that is apt for the local context. Ethically, the important outcomes (regardless of the policy chosen) might include that burdens, including risks (to both participants and third parties), are appropriately minimised; that public trust in research is maintained; and that scientifically valuable, acceptably low-risk studies are not unduly impeded by excessively costly or slow review procedures (Eyal et al. 2018).

As noted in other sections of this report, public controversies have the potential to undermine support/acceptance of (other) research and/or public health endeavours—domestically and/or internationally. Thus, in some cases, international consultation and/or appeal to international agencies (e.g., WHO) may be appropriate. Whether or not a particular jurisdiction decides on specialised or standard review, international agreement on an ethical and regulatory framework for HCS may help to improve review and ensure that relevant issues are consistently addressed (WHO Expert Committee on Biological Standardization 2016). Stakeholders interviewed in this project held a range of different views, with some in favour of standard review procedures (see Box 4.3 above) whereas others favoured some form of specialised review (see Box 4.5). Finally, some raised potential problems with specialised review (see Box 4.6).

Box 4.5 Advantages of specialised ethics review for HCS

IRBs work really well for fairly routine … garden variety research, but I think when you're at the vanguard, you want some sort of specialised mechanism and … I think there are substantive reasons for that, that get to the quality of the expertise that you get – and specialised review, the ability to pick through and second guess the scientific rationale, is really key. [Jonathan Kimmelman, ethicist, Canada]

I think the most important thing is to be transparent. So whatever process you set up – and perhaps having a central high-level review mechanisms for all such studies is the way to go – … to make it clear that there is nothing being hidden from anybody is

---

the important thing. To some extent, having a centrally mandated committee would also be helpful because it would provide some distancing and protection from the investigators and their institutions. [Gagandeep Kang, scientist, India]

[T]here may be some cases where having an extra layer of review that the researcher either voluntary agrees to or that the sponsor puts in a place can help make sure that everything is done as rigorously and carefully as possible and then … reassure the existing levels of review that we already have. [Ethicist, North America]

Box 4.6 Potential problems with special review for HCS

[S]pecial challenge ethics committees … can give you clearance to do stuff which nobody else can and [that] doesn't sit right with me … [I]t should be … the general ethics committee and … they should have training and they should understand the issues and … of course, also I think investigators [should] make sure that the issues are clearly articulated. [Scientist, Asia]

I suppose you could [have a specialised IRB for HCS] but they're so rare in a given place … and to set up something separate just for one study a year for challenge [studies] is a bit over the top I feel. It's an over-response to a problem that doesn't exist. [Scientist, North America]

I'm not sure that having a panel of experts will speed anything up! … [W]e had difficulty enough just explaining the host country [i.e., LMIC] processes to [those involved in] the US regulatory review process … I would think that kind of requirement should probably come from the host country. If the host country wants to have [an] additional advisory [body] or ask the WHO, or ask some other group … it's really up to them to decide what level of review they think is necessary. [Carl Mason, scientist, USA]

4.3 Regulation

HCS are governed by standard regulations related to the scientific conduct of research (including the need for ethical review procedures, etc.) and those related to the development and use of investigational interventions (e.g., where vaccines or drugs are tested during HCS). More specifically, HCS may be subject to particular regulations related to the development and use of a challenge organism (and, in some cases, additional regulations if the organism is genetically modified) (Academy of Medical Sciences 2018). Since the general regulations governing research and the use of investigational vaccines are not specific to HCS, this section focuses particularly on (i) the regulation of challenge organisms, and (ii) the role of HCS in regulatory development pathways towards the licensure of new vaccines (and/or treatments).

4.3.1 International Regulations

In 2016, the WHO Expert Committee on Biological Standardisation published a short guidance document entitled 'Human Challenge Trials for Vaccine Development: Regulatory Considerations' (WHO Expert Committee on Biological Standardization 2016). The Working Group involved in the preparation of this document included representatives from HIC regulators and research institutes as well as similar bodies in Sub-Saharan Africa. The guidance document does not contain binding requirements but is intended to provide general advice to regulators and manufacturers of biological products in WHO Member States. Overall, it is suggested that HCS should be conducted in a similar way to a (non-HCS) vaccine study—i.e., following standard requirements for clinical research (e.g., International Council for Harmonisation of Technical Requirements for Pharmaceuticals for Human Use (ICH) Good Clinical Practice (CGP) procedures and local regulatory Clinical Trial Authorisation (CTA) procedures for the conduct of a study)—although the document notes potential variations in local regulations, for example those governing genetically modified organisms.

Among other points, the guidance document notes that HCS have been conducted in LMICs, and that, where this occurs, "the same standards apply as in more developed countries" (WHO Expert Committee on Biological Standardization 2016), including compliance with local regulations (with a recommendation for the establishment of an appropriate framework for challenge studies if none exists), and ethics review with a particular emphasis on risks to participants and third parties. Further, the document recommended that ethical considerations regarding HCS should be thoroughly evaluated. As of 2019, WHO has initiated a process to review relevant considerations with the goal of developing ethical guidance related to HCS.

4.3.2 Regulating Challenge Strains

The majority of HCS have been conducted in HICs, particularly in the UK/Europe and USA. Challenge studies in LMICs have also frequently involved collaborators from HICs and, with the exception of the Colombian vivax HCS program (see Chap. 5), it has usually been the case that the challenge organism has undergone part or all of its development in an HIC before being used in an LMIC HCS (although this may change with future capacity building). Nevertheless, many challenge strains originated in LMICs (particularly for pathogens primarily endemic to LMICs): for example, although the NF54 malaria parasite used in the African challenge studies was first obtained from a patient in the Netherlands, and subsequently developed primarily in the USA, there is evidence that it originated in Africa (Eldering et al. 2016).

In cases where a strain (wherever it originated) undergoes preparation in a HIC before use in an LMIC HCS, the challenge organism will usually be subject to regulatory oversight in both countries—as occurred for the African studies reviewed below (which involved a strain prepared in the USA). Thus, we review US, UK, and European regulatory requirements below before discussing the regulation of challenge strains in LMICs.

4.3.2.1 US, UK and European Regulatory Requirements

The US Food and Drug Administration (FDA) requires that challenge organisms comply with current Good Manufacturing Practices (cGMP) regulations, a set of controls that aim to ensure the safe production, monitoring, and use of investigational agents (including drugs, vaccines, and challenge organisms) for use in humans.[1] The situation for genetically modified challenge strains (which, to date, have not been used in LMIC HCS) is more complex, requiring review by the local USA Institutional Biosafety Committee (IBC), and—in some cases—the Recombinant DNA Advisory Committee (RAC). Of particular importance for LMIC HCS, the FDA also has jurisdiction over challenge strains exported from the USA to other countries (with the exception of a short list of countries with national regulatory authorities (NRAs) recognised by the FDA[2]) (Academy of Medical Sciences 2018; Baay et al. 2018).

The US FDA is relatively unique in its regulation of challenge strains—with the norm in many other jurisdictions being more ad hoc and/or responsibility for the quality of a challenge strain devolving to the (usually academic) institution where the organism is prepared. As the 2016 WHO document on regulations governing challenge studies noted, "in many countries, because the challenge stock … itself is not considered to be a medicinal product, [challenge studies without the use of an investigational drug or vaccine] would not come under the NRA's review and authorization. Thus, much less clarity exists on regulatory expectations and quality matters in such cases" (WHO Expert Committee on Biological Standardization 2016).

For example, the UK Medicines and Healthcare products Regulatory Agency (MHRA) does not currently require that challenge organisms are prepared according to the kind of stringent regulations applied to vaccines and drugs, nor does it require a CTA for a particular HCS unless it also employs an investigational drug or vaccine (e.g., to be tested against an infection challenge). This largely reflects the international "default" where there are no specific regulations for challenge strains (perhaps because of the relative novelty of HCS and small number of sites conducting such studies). Thus, in the UK and in many other countries,

---

[1]See   https://www.fda.gov/drugs/developmentapprovalprocess/manufacturing/ucm169105.htm. [Accessed 29 March 2019].

[2]EU/EEC, Australia, Canada, Japan, Israel, New Zealand, Switzerland, South Africa—see FDA 21 CFR 312.110.

Table 4.1 Regulatory bodies and/or specific regulations relevant to challenge organisms

International

Regulatory agency	Relevant regulation	Comments
WHO (ECBS)	Human challenge trials for vaccine development: regulatory considerations (2016)	General advisory document, no specific requirements for challenge strain. Recommended for implementation in WHO member states

National/regional (HIC)

Country/region regulatory agency	Relevant regulation	Comments
United Kingdom MHRA	Nil specific challenge strain regulation	ACRE and DEFRA (if genetically modified)
United States of America US FDA	cGMP (for both challenge strain and other investigational products)	plus IBC and/or RAC (if genetically modified)
Europe European Medicines Agency (EMA)	Auxiliary medicinal products in clinical trials (2017)	EMA guidance is non-binding and requires member state ratification

ECBS Expert Committee on Biological Standardisation, *FDA* Food and Drug Administration, *cGMP* current Good Manufacturing Practices, *IND* Investigational New Drugs, *IBC* Institutional Biosafety Committee, *RAC* Recombinant DNA Advisory Committee, *ACRE* UK Advisory Committee on Releases to the Environment, *DEFRA* Department for Environment, Food, and Rural Affairs

HCS are conducted largely within academic institutions and these institutions are de facto responsible for the development and use of challenge strains (see Box 4.7). For genetically modified organisms (GMOs), additional requirements apply, similar to those in the USA. For example, the UK Advisory Committee on Releases to the Environment (ACRE) and the Department for Environment, Food, and Rural Affairs (DEFRA) have (sometimes-overlapping) responsibilities for GMOs (whether or not scientists intend to release such a challenge strain beyond the laboratory) (Academy of Medical Sciences 2018) (see Table 4.1).

Box 4.7 Regulating challenge strains

[In the UK, challenge strains such as] malaria parasites are not a regulatory product. So we have ethical approval but no regulatory approval, and so the stringent assessments of the quality of products and storage and all those other things, we just don't have anyone who controls that and so you don't have that added reassurance that you might do with the GMP. That's one thing that could come in that could really help control the field and give some reassurance about the whole process. [Scientist, UK/Europe]

When we're talking about the challenge strains, the manufacturing quality, purity, and reproducibility from batch to batch... all of these things are really important considerations but they're not, at least in the UK, currently subjected to the same level of regulatory scrutiny [as, for example, vaccines]. So it's a little bit of a grey area, and it benefits us because it makes it easier to do the challenge studies but also there's potential risk there, I think, and maybe it is touching some ethical issues as well because actually you know, we make all of our challenge strains to a GMP standard; it's very time consuming, very costly, but it has benefits: a fair amount of stability, reproducibility from batch to batch, and also we can have a degree of confidence that we're happy with what is given to our volunteers. [The volunteers] assume that what you're challenging them with is going to be safe. [Malick Gibani, scientist, UK]

I think from a regulatory point of view they do need special handling because there are issues like understanding a non-GMP manufacturing process that is still safe enough—because [not all challenge strains are] manufacturable to GMP. Probably those [GMP] standards are excessive for something which is never going to be a final product. You can have very tight systems around producing a challenge agent but [that] are not as [extensive] as GMP, and defining what those are, I think, is difficult to do in the abstract; you have to talk about individual agents... but I think, from a regulation point of view, there has to be enormous expertise in looking at those agents to make sure that they are as safe as possible and that protocols include information to show the limitations of that. [Scientist, UK/Europe]

[W]e could reconsider how we're requiring the production and the characterisation of the challenge product because that sometimes ... is limiting because ... if there's only so many products that have been made in this GMP manufacturing style, you're limited [with respect] to the study questions that you can ask and the relevance of that. And ... it's limiting in a way because of that huge amount of scrutiny/oversight of that product. At the same time, you want to make sure that you're delivering a product that's safe. [Scientist, North America]

As an investigator I would say [that regulating challenge strains is] not necessary or I would say it's very cumbersome. But, on the other hand, it is also very important that there is an independent review of the GMP aspects of the challenge strains because ... there is risk. Maybe it's not so high, but there is quite a bit of risk. But I think it's a good [mechanism] at the European level, what they call [an] "auxiliary medicinal product" [because] the set-up of preclinical investigations may be quite different for something that you only use ... for such a purpose in clinical [challenge] trials, compared to something [like a vaccine] that you readily release [for use in] the health system. [Benjamin Mordmüller, scientist, Germany]

The European Union, via the European Medicines Agency (EMA), does have GMP requirements similar to those of the FDA—but these procedures typically apply to *facilities* that produce agents for human use, rather than to individual *products*. However, each product must have a Qualified Person who is uniquely responsible for each and every release of a clinical trial lot of material for research. These requirements potentially extend to challenge organisms—which are covered by EMA regulations as Auxiliary Medicinal Products (AMPs); however, these regulations do not include detailed requirements for infection challenge strains in particular and, in any case, all EMA regulations must be implemented by each EU

Member State and (as of early 2019) they have not yet been widely ratified (Baay et al. 2018).

Some pathogens that cannot currently be maintained in a laboratory (e.g., vivax malaria) may not be able to be prepared according to strict GMP requirements. Several investigators interviewed for this project indicated that they nevertheless routinely follow GMP requirements to the extent that this is possible within current scientific limits. Overall, many stakeholders felt that clear and (ideally internationally) consistent guidance from regulatory bodies would be useful to researchers, although there were unresolved issues regarding the optimum regulatory model for challenge strains and for HCS more generally (see Box 4.7 above). With recent increases in HCS in both HICs and LMICs, there will be opportunities for regulatory agencies to develop appropriate national, regional, and international norms specific to challenge strains and other aspects of HCS, including their role in licensure of new interventions (discussed in Sect. 4.3.3).

4.3.2.2   Low- and Middle-Income Country Regulatory Requirements

Given the relatively small number of challenge studies performed in LMICs to date, many LMICs may not yet have specific regulations related to HCS and/or challenge strains—although, similar to countries like the UK, there are usually local regulations governing investigational drugs and vaccines (which may be used in some HCS). Nevertheless, challenge strains have sometimes been reviewed and approved for use in research by relevant LMIC pharmaceutical regulators (see Table 4.2), generally after approval by a HIC regulator (e.g., US FDA) for international collaborative studies (Hodgson et al. 2014, 2015; Shekalaghe et al. 2014). Likewise, researchers in some countries are required to obtain clearance from other agencies (generally Ministry of Health or similar) for certain types of

Table 4.2 LMIC regulators

Country	Regulator	Approval of challenge strains	Other relevant agencies
Colombia	Instituto Nacional de Vigilancia de Medicamento	N/A	
Gabon	Direction Médicament et de la Pharmacie	Yes	Ministry of Health
Kenya	Pharmacy and Poisons Board	Yes	
Tanzania	Tanzanian FDA	Yes	
Thailand	Thai FDA	N/A	Ministry of Public Health

Note No specific regulations govern challenge strains in these jurisdictions; African regulators listed above have approved challenge strains in conjunction with US FDA research (e.g., research with potential public health implications such as third-party risks).

This pattern of prior approval of a challenge strain in a HIC prior to use in LMICs may be due to a combination of multiple factors (see Box 4.8), including (i) lack of laboratory capacity for the development of challenge strains in many LMICs, (ii) lack of regulatory capacity for thorough review of challenge strains (meaning that LMIC regulators may defer to HIC regulatory review) in part due to the low number of HCS so far conducted in LMICs, and/or (iii) de facto or more formal requirements from LMIC regulators and/or other institutions that HIC sponsors test investigational agents (whether vaccines, drugs, or challenge strains) in HIC populations prior to testing in LMICs (see Box 4.8).

Box 4.8 Regulatory norms and practices in LMICs

[S]ome regulators may not be comfortable taking a model into their country until it's been used in the US, and that can be for a number of reasons. [First, regulators might not like] their people being experimented on as 'guinea pigs', and so they want data from the US first. [Second,] some of it is just this idea ... that if something goes first to an endemic setting it's ethically suspect and it's something that wouldn't have been allowed in the US, and so it's like an ethical double standard. [Third,] some of it may be just the people want some preliminary data before they're comfortable with allowing a study to go forward. But whatever it is ... I've noticed that sometimes there is this resistance by regulators to having a study be conducted in their [LMIC] setting if it hasn't first been done in a [HIC] non-endemic setting. [Ethicist, North America]

[O]n the GMP side of things, how this is produced and how [it] is ensured that no contaminations are there, and release criteria, [and so on, are things] that we have not discussed with the regulators in Gabon ... [I]t is also because these assessors there, they are not ... trained like those at the FDA for example ... but there are now programs to improve this regulatory environment and capacities in Africa ... I know there's one, for example, between these German regulators and these African regulators, because they've seen that this is a problem and they have to rely really on the judgment of others. [Benjamin Mordmüller, scientist, Germany]

[There is a] concern that developing countries don't have the regulatory infrastructure to really evaluate these studies and they don't have the clinical trial [infrastructure] or [other important] infrastructure to do these studies. So [LMIC regulators] want to make sure that, before a challenge study goes to an endemic area, it's been thoroughly [tested] in the United States, because it gives regulatory authorities in the developing countries confidence that they're not the first to see this. I'm hoping that as we build up the regulatory infrastructure in some of these developing countries, that they will feel more confident in being first, but I don't think they're there yet and... I don't necessarily think challenge studies have to be done in developed countries first. [Anna Durbin, scientist, USA]

Some of the regulators in Malawi [said] 'What regulations do you have?' and we [said] 'Oh in the UK we have this, and in the US they have that and in France they have this' and so, for setting it up, the regulatory considerations are really important in an endemic setting but there's nothing that you can copy and paste; there's no

> gold standard that you can refer to, and that's a potential problem, and you know,
> will the government of Uganda decide to take a liberal view on challenge studies
> whereas the government [of] Malawi decide to take a more conservative view on
> that? [It would help to have] a bit more clarity and … consistency [internationally].
> [Dr. Malick Gibani, scientist, UK]

Stakeholders interviewed in this project identified regulation of research (and HCS in particular) in LMICs as an area that might benefit from capacity building. Fortunately, there are a number of on-going LMIC regulatory capacity building initiatives, particularly in Africa (Ndomondo-Sigonda et al. 2017), although these are usually focused on regulation of vaccines and therapeutics, rather than challenge strains. Likewise, there are opportunities for communication between HIC and LMIC regulators, although these tend to be ad hoc rather than usual practice according to formalised procedures (personal communication, expert stakeholder). This review focused on the five LMICs in which HCS have been conducted and published since 1992; a wider review of the regulatory environment in other countries considering or conducting HCS could help to inform future regulatory guidance and international collaborations.

4.3.3 Challenge Studies and Licensure of New Interventions

A key unresolved question related to regulatory approval of new interventions (e.g., vaccines) is the role of HCS in the development pathway towards approval/licensure. HCS can play a role at multiple steps in development pathways, for example in the context of (i) basic science and very early phase research (e.g., exploring infectious disease pathogenesis, immune correlates of protection, and developing models of infection), (ii) phase IIB studies (i.e., providing preliminary estimates of efficacy to enable the selection of candidate interventions for prioritised investigation in field trials), and (iii) phase III studies (i.e., more definitive estimates of efficacy intended to obviate the need for field trials) (Sauerwein et al. 2011; Chattopadhyay and Pratt 2017; Shah et al. 2017; Baay et al. 2018; Roestenberg et al. 2018a).

The degree to which phase 2B HCS can be used to predict (phase 3) field trial efficacy (which is sometimes sufficient to support licensure of a vaccine) and, similarly, the degree to which phase 3 HCS can support licensure decisions will depend in part on whether the findings of the HCS are generalisable to the disease epidemiology in the target population for the intervention (see Sect. 3.2.1.1). For example, (i) results from HCS in US volunteers were used to support licensure of a cholera vaccine for travellers (but not for those living in endemic settings because the results were not considered generalisable to these populations—in whom higher doses of the vaccine were required to generate a similar immune response

(Sow et al. 2017))[3] and (ii) results from HCS in UK volunteers were used to support WHO prequalification (which endorses the product for licensure in WHO member states) of the first typhoid vaccine (which is manufactured by an Indian company) (Jin et al. 2017; World Health Organization 2018). The latter HCS used a wild-type typhoid strain (making it more generalisable to natural infection than attenuated strains) and HCS results with this strain were found to correlate with immunogenicity in (non-HCS) field trials in children in endemic areas, thus supporting using this vaccine in such settings/populations (Feasey and Levine 2017; Jin et al. 2017).

HCS designs have also been used to provide preliminary (Phase 2B) estimates of vaccine efficacy for malaria vaccines that have subsequently entered clinical trials (e.g., the PfSPZ vaccine (Hoffman et al. 2002; Olotu et al. 2018)) and/or been licensed for use (e.g., the RTS,S vaccine (Ballou 2009; RTS 2015)) and to 'deselect' vaccines that showed no efficacy against malaria challenge (Spring et al. 2009; Roestenberg et al. 2018b) including in a Gabonese HCS reviewed below (Dejon-Agobe et al. 2018). Safety and/or efficacy trials of at least two malaria vaccines using HCS designs have taken place in Sub-Saharan African countries (Dejon-Agobe et al. 2018; Olotu et al. 2018). The degree to which current falciparum malaria HCS designs can be used to supplant field trials, however, remains controversial. This is in part because few falciparum malaria strains are available for use in HCS, and some doubt that efficacy measured against this limited number of strains would be generalisable to diverse malaria infections "in the wild" (see Box 4.9 and also Box 3.3 in Sect. 3.2.1.1) (Chattopadhyay and Pratt 2017).

Box 4.9 The role of HCS in licensure of new interventions

[A] couple of years ago [in 2016] we approved a cholera vaccine [for travellers to endemic countries] where the efficacy data was based entirely on a human challenge study. So that sort of creates a new paradigm or it establishes a precedent for a new paradigm. [Although] this is not something FDA did in isolation. We had … public advisory committee meetings to discuss this pathway toward approval … and there was agreement from our [external] advisory committee [made up of infectious disease experts with no declared conflicts of interest] that this would be a reasonable way to proceed. [Regulatory representative, North America]

[T]he regulatory authorities as far as I can see … tend to be extremely wedded to the way they've always done things, so I think somebody should, people should, be making … the case for looking [at the role of HCS in licensure pathways] from first principles and looking at what is the evidence that this is likely to predict future benefit. [Scientist, UK/Europe]

[F]or malaria the problem is: Can you licence a product on the basis of a few human challenges done with a few different strains and say that that provides coverage for

---

[3]For details of FDA licensure, see https://www.fda.gov/downloads/BiologicsBloodVaccines/Vaccines/ApprovedProducts/UCM509681.pdf. [Accessed 1 March 2019].

> the diversity of parasites out there in the world? That's the biggest regulatory hurdle … [North American Scientist]
>
> I think a clearer, better lined regulatory pathway would be helpful. WHO does not rule the regulators of … the European Union, the U.K., Japan or the U.S. but certainly does influence very heavily the regulators in low- and middle-income countries. It would be nice to see more from WHO [regarding] how such studies should be done. [Gagandeep Kang, scientist, India]

4.3.4 Regulation of Over-Volunteering

Since HCS often attract high levels of payment, they may be one area of research that could be particularly likely to be undermined, scientifically and ethically, by over-volunteering. The underlying ethical concerns are that participation in too many studies, too frequently, may (i) lead to excessive risk to volunteers and/or (ii) distort research results (or interpretation thereof) when a participant is not representative of the general/study population due to past interventions, possibly including relevant vaccines and/or other challenge strains. Some interviewees identified concerns related to over-volunteering, particularly in LMICs (see Sect. 3.6.2); however, few empirical data are available to assess the extent and effects of this phenomenon in particular countries.

In the UK, The Over-volunteering Prevention System (Allen et al. 2017), governed by the MHRA, requires that participants enrolled in (predominantly phase I) studies of healthy volunteers are registered to prevent overly-frequent research participation. It is unclear whether HCS participants are always required to be enrolled in this system. Similar systems exist in France, Germany, the Netherlands, Belgium, and Switzerland, although to our knowledge these systems are optional rather than required. In any case, it is widespread common practice to exclude those known to have recently participated in other research involving exposures that could interfere with the results of subsequent studies. In the absence of a formal regulatory system, this relies on potential HCS participants to declare such information—but they may be especially reluctant to do so in cases where study participation is motivated by monetary payment. Thus, this may be an area in which other national regulators and/or policymakers may be able develop systems that reduce risks to participants and support safe, efficient, and transparent research with healthy volunteers.

4.3.5 Laws Criminalising Intentional Infection

In the background of specific regulations governing research, some countries may also have laws prohibiting the intentional infection of individuals with pathogens. For example, the UK only repealed a law of this kind in 2010 (Brazier and Gostin 2016). Thus, researchers and regulators in each country must also be mindful of local laws (beyond specific research regulations) that may be relevant to HCS, including those that criminalise infecting others. Depending on how such laws are interpreted, they may, in some cases, preclude conducting HCs in a particular country. There are on-going debates regarding the ethics of criminalising infectious disease transmission (Stanton and Quirk 2016) and relevant future work in this area could include a review of international laws regarding infectious diseases, especially in countries considering conducting (more) HCS.

References

Academy of Medical Sciences. 2018. *Controlled human infection model studies: Summary of a workshop held on 6 February 2018*. London: Academy of Medical Sciences.

Allen, C., G. Francis, J. Martin, and M. Boyce. 2017. Regulatory experience of TOPS: An internet-based system to prevent healthy subjects from over-volunteering for UK clinical trials. *European Journal of Clinical Pharmacology* 73 (12): 1551–1555.

Baay, M.F.D., T.L. Richie, P. Neels, M. Cavaleri, R. Chilengi, D. Diemert, S.L. Hoffman, R. Johnson, B.D. Kirkpatrick, and I. Knezevic. 2018. Human challenge trials in vaccine development, Rockville, MD, USA, September 28–30, 2017. *Biologicals*.

Ballou, W.R. 2009. The development of the RTS, S malaria vaccine candidate: Challenges and lessons. *Parasite Immunology* 31 (9): 492–500.

Bambery, B., M. Selgelid, C. Weijer, J. Savulescu, and A.J. Pollard. 2015. Ethical criteria for human challenge studies in infectious diseases. *Public Health Ethics* 9 (1): 92–103.

Brazier, M., and L.O. Gostin. 2016. Foreword. In *Criminalising contagion: Legal and ethical challenges of disease transmission and the criminal law*, ed. C. Stanton, and H. Quirk. Cambridge University Press.

Chattopadhyay, R., and D. Pratt. 2017. Role of controlled human malaria infection (CHMI) in malaria vaccine development: A US food and drug administration (FDA) perspective. *Vaccine* 35 (21): 2767.

Davies, H. 2019. Human challenge studies, topic entry in 'Reviewing research'. http://www.reviewingresearch.com/human-challenge-studies/. Accessed 30 Mar 2019.

Dejon-Agobe, J.C., U. Ateba-Ngoa, A. Lalremruata, A. Homoet, J. Engelhorn, O. Paterne Nouatin, J.R. Edoa, J.F. Fernandes, M. Esen, and Y.D. Mouwenda. 2018. Controlled human malaria infection of healthy lifelong malaria-exposed adults to assess safety, immunogenicity and efficacy of the asexual blood stage malaria vaccine candidate GMZ2. *Clinical Infectious Diseases*.

El Setouhy, M., T. Agbenyega, F. Anto, C.A. Clerk, K.A. Koram, M. English, R. Juma, C. Molyneux, N. Peshu, and N. Kumwenda. 2004. Moral standards for research in developing countries from "Reasonable Availability" to "Fair Benefits". *The Hastings Center Report* 34 (3): 17–27.

Eldering, M., I. Morlais, G.-J. van Gemert, M. van de Vegte-Bolmer, W. Graumans, R. Siebelink-Stoter, M. Vos, L. Abate, W. Roeffen, and T. Bousema. 2016. Variation in susceptibility of African Plasmodium falciparum malaria parasites to TEP1 mediated killing in Anopheles gambiae mosquitoes. *Scientific Reports* 6: 20440.

Eyal, N., M. Lipsitch, T. Bärnighausen, and D. Wikler. 2018. Opinion: Risk to study nonparticipants: A procedural approach. *Proceedings of the National Academy of Sciences* 115 (32): 8051–8053.

Feasey, N.A., and M.M. Levine. 2017. Typhoid vaccine development with a human challenge model. *The Lancet* 390 (10111): 2419–2421.

Gikonyo, C., P. Bejon, V. Marsh, and S. Molyneux. 2008. Taking social relationships seriously: Lessons learned from the informed consent practices of a vaccine trial on the Kenyan Coast. *Social Science and Medicine* 67 (5): 708–720.

Hodgson, S.H., E. Juma, A. Salim, C. Magiri, D. Kimani, D. Njenga, A. Muia, A.O. Cole, C. Ogwang, and K. Awuondo. 2014. Evaluating controlled human malaria infection in Kenyan adults with varying degrees of prior exposure to Plasmodium falciparum using sporozoites administered by intramuscular injection. *Frontiers in Microbiology* 5: 686.

Hodgson, S.H., E. Juma, A. Salim, C. Magiri, D. Njenga, S. Molyneux, P. Njuguna, K. Awuondo, B. Lowe, and P.F. Billingsley. 2015. Lessons learnt from the first controlled human malaria infection study conducted in Nairobi, Kenya. *Malaria Journal* 14 (1): 182.

Hoffman, S.L., L.M.L. Goh, T.C. Luke, I. Schneider, T.P. Le, D.L. Doolan, J. Sacci, P. De la Vega, M. Dowler, and C. Paul. 2002. Protection of humans against malaria by immunization with radiation-attenuated Plasmodium falciparum sporozoites. *The Journal of Infectious Diseases* 185 (8): 1155–1164.

Hope, T., and J. McMillan. 2004. Challenge studies of human volunteers: Ethical issues. *Journal of Medical Ethics* 30 (1): 110–116.

Jin, C., M.M. Gibani, M. Moore, H.B. Juel, E. Jones, J. Meiring, V. Harris, J. Gardner, A. Nebykova, and S.A. Kerridge. 2017. Efficacy and immunogenicity of a Vi-tetanus toxoid conjugate vaccine in the prevention of typhoid fever using a controlled human infection model of Salmonella Typhi: A randomised controlled, phase 2b trial. *The Lancet*.

Miller, F.G., and C. Grady. 2001. The ethical challenge of infection-inducing challenge experiments. *Clinical Infectious Diseases* 33 (7): 1028–1033.

Ndomondo-Sigonda, M., J. Miot, S. Naidoo, A. Dodoo, and E. Kaale. 2017. Medicines regulation in Africa: Current state and opportunities. *Pharmaceutical Medicine* 31 (6): 383–397.

Njue, M., P. Njuguna, M.C. Kapulu, G. Sanga, P. Bejon, V. Marsh, S. Molyneux, and D. Kamuya. 2018. Ethical considerations in controlled human malaria infection studies in low resource settings: Experiences and perceptions of study participants in a malaria challenge study in Kenya. *Wellcome Open Research* 3.

Njue, M., F. Kombe, S. Mwalukore, S. Molyneux, and V. Marsh. 2014. What are fair study benefits in international health research? Consulting community members in Kenya. *PLoS ONE* 9 (12): e113112.

Olotu, A., V. Urbano, A. Hamad, M. Eka, M. Chemba, E. Nyakarungu, J. Raso, E. Eburi, D.O. Mandumbi, and D. Hergott. 2018. Advancing global health through development and clinical trials partnerships: A randomized, placebo-controlled, double-blind assessment of safety, tolerability, and immunogenicity of PfSPZ vaccine for malaria in healthy equatoguinean men. *The American Journal of Tropical Medicine and Hygiene* 98 (1): 308–318.

Roestenberg, M., I.M.C. Kamerling, and S.J. de Visser. 2018a. Controlled human infections as a tool to reduce uncertainty in clinical vaccine development. *Frontiers in Medicine* 5.

Roestenberg, M., I. Kamerling, and S.J. de Visser. 2018b. Dealing with uncertainty in vaccine development: The malaria case. *Frontiers in Medicine* 5: 297.

Rts, S. 2015. Efficacy and safety of RTS, S/AS01 malaria vaccine with or without a booster dose in infants and children in Africa: Final results of a phase 3, individually randomised, controlled trial. *The Lancet* 386 (9988): 31–45.

Sauerwein, R.W., M. Roestenberg, and V.S. Moorthy. 2011. Experimental human challenge infections can accelerate clinical malaria vaccine development. *Nature Reviews Immunology* 11 (1): 57.

Selgelid, M. 2013. The ethics of human microbial challenge (conference paper). In *Controlled human infection studies in the development of vaccines and therapeutics*. Jesus College, Cambridge, UK.

Shah, S.K., J. Kimmelman, A.D. Lyerly, H.F. Lynch, F. McCutchan, F.G. Miller, R. Palacios, C. Pardo-Villamizar, and C. Zorilla. 2017. Ethical considerations for Zika virus human challenge trials. *National Institute of Allergy and Infectious Diseases*.

Shah, S.K., J. Kimmelman, A.D. Lyerly, H.F. Lynch, F.G. Miller, R. Palacios, C.A. Pardo, and C. Zorrilla. 2018. Bystander risk, social value, and ethics of human research. *Science* 360 (6385): 158–159.

Shekalaghe, S., M. Rutaihwa, P.F. Billingsley, M. Chemba, C.A. Daubenberger, E.R. James, M. Mpina, O.A. Juma, T. Schindler, and E. Huber. 2014. Controlled human malaria infection of Tanzanians by intradermal injection of aseptic, purified, cryopreserved Plasmodium falciparum sporozoites. *The American Journal of Tropical Medicine and Hygiene* 91 (3): 471–480.

Sow, S.O., M.D. Tapia, W.H. Chen, F.C. Haidara, K.L. Kotloff, M.F. Pasetti, W.C. Blackwelder, A. Traoré, B. Tamboura, and M. Doumbia. 2017. Randomized, placebo-controlled, double-blind phase 2 trial comparing the reactogenicity and immunogenicity of a single standard dose to those of a high dose of CVD 103-HgR live attenuated oral cholera vaccine, with Shanchol inactivated oral vaccine as an open-label immunologic comparator. *Clinical Vaccine Immunology* 24 (12): e00265–00217.

Spring, M.D., J.F. Cummings, C.F. Ockenhouse, S. Dutta, R. Reidler, E. Angov, E. Bergmann-Leitner, V.A. Stewart, S. Bittner, and L. Juompan. 2009. Phase 1/2a study of the malaria vaccine candidate apical membrane antigen-1 (AMA-1) administered in adjuvant system AS01B or AS02A. *PLoS ONE* 4 (4): e5254.

Stanton, C., and H. Quirk. 2016. *Criminalising contagion: Legal and ethical challenges of disease transmission and the criminal law.* Cambridge University Press.

UK Academy of Medical Sciences. 2005. *Microbial challenge studies of human volunteers.* London: Academy of Medical Sciences.

WHO Expert Committee on Biological Standardization. 2016. *Human challenge trials for vaccine development: Regulatory considerations.* Geneva: World Health Organization.

World Health Organization. 2017. *Ethical issues associated with vector-borne diseases: Report of a WHO Scoping Meeting.* Geneva: WHO.

World Health Organization. 2018. Typhoid vaccien prequalified. https://www.who.int/medicines/news/2017/WHOprequalifies-breakthrough-typhoid-vaccine/en/. Accessed 1 Mar 2019.

Chapter 5
Case Studies: Challenge Studies in Low- and Middle-Income Countries

This review identified 13 case studies involving primary publications detailing HCS in 5 LMICs published from 1992–2018:

(i) 4 enteric pathogen HCS in Thailand with cholera (Suntharasamai et al. 1992; Pitisuttithum et al. 2002) and *Shigella* (Bodhidatta et al. 2012; Pitisuttithum et al. 2016),

(ii) 5 falciparum malaria HCS in Sub-Saharan Africa in Tanzania (Shekalaghe et al. 2014; Jongo et al. 2018), Kenya (Hodgson et al. 2014), and Gabon (Lell et al. 2017; Dejon-Agobe et al. 2018), and

(iii) 4 vivax malaria HCS in Colombia (Herrera et al. 2009; Herrera et al. 2011; Arévalo-Herrera et al. 2014; Vallejo et al. 2016).

These endemic-region HCS together recruited approximately 400 participants—which, as mentioned earlier, amounts to less than 1% of the >40,000 human volunteers who have participated in HCS worldwide (i.e., >99% of HCS participants have been in HICs) since World War II (Evers et al. 2015; Kalil et al. 2012). While other LMIC HCS may have taken place during this period of time (e.g., unpublished studies and/or those not captured in the detailed case studies below), the authors who estimated that 40,000 individuals have participated in HCS overall suggested that this total would be an underestimate—meaning that LMIC participants have in any case been grossly under-represented (Evers et al. 2015). This suggests that endemic-region HCS has been a neglected area of research, even though the vast majority of infectious disease morbidity and mortality occurs in endemic LMICs. In addition to the above studies, HCS are currently being considered and/or conducted in LMICs including Equatorial Guinea, Gabon, India, Indonesia, Kenya, Malawi, Mali, Tanzania, Thailand, Uganda, and Vietnam (Personal communications from study participants) (Baay et al. 2018).

Although these 13 published studies all took place within endemic countries that have active transmission of the pathogen in question in at least part of the country, they frequently took place in a city/location where there was no local transmission (e.g., the studies in Kenya and Colombia were in non-endemic areas). We discuss the ethical salience of decisions regarding the location of study sites below.

© The Author(s) 2021

E. Jamrozik and M. J. Selgelid, *Human Challenge Studies in Endemic Settings*, SpringerBriefs in Ethics, https://doi.org/10.1007/978-3-030-41480-1_5

These 13 studies from 1992 to 2018 were increasingly pre-registered (in line with general trends for clinical research (Ioannidis 2015)). Pre-registration is arguably ethically important since it can help to (i) improve transparency (e.g., by requiring analyses and methods to be specified in advance rather than altered once the trial is in progress), (ii) reduce publication bias (e.g., by providing an incentive for publication of all research findings, whether favourable or not), (iii) reduce unnecessary duplication of research efforts (thus reducing the chance that participants will be exposed to challenge infections unnecessarily) and (iv) increase standardisation and/or comparability across similar research programs (if research groups are able to co-ordinate) which may serve to increase the accuracy and impact of results (Ioannidis 2015). As one interviewee argued:

> [It's important that] the results are disseminated very widely, whatever the results are, which is a major issue that we have across all of research … you're putting these individuals at more risk than in some clinical research, it makes it get an even greater imperative to disseminate the results, totally transparently. [Scientist, UK/Europe]

5.1 Cholera and *Shigella* Challenge Studies in Thailand

We identified 4 studies of diarrhoeal disease using HCS designs in Thailand, all using models developed in previous (non-endemic) North American studies. These comprised 2 HCS that aimed to replicate infection models of cholera (*Vibrio cholerae*) developed in the US (Suntharasamai et al. 1992; Pitisuttithum et al. 2002); and 2 HCS using *Shigella sonnei* (a major cause of bacterial dysentery), first to develop an infection model (Bodhidatta et al. 2012) and subsequently to test a live attenuated vaccine using this model (Pitisuttithum et al. 2016) (See Table 5.1).

5.1.1 Rationale and Review Process

The cholera HCS conducted at Mahidol Univeristy, Bangkok, Thailand, published by Suntharasamai et al. (1992) is, to our knowledge, the first endemic-region LMIC HCS since 1956 (see history section above) (Suntharasamai et al. 1992) and was followed by a similar HCS by the same group with a different cholera serotype in 2002 (Pitisuttithum et al. 2002). The stated rationale for the two cholera studies was to replicate previous cholera HCS studies (in non-endemic populations) because "ethnic factors, gut flora, and immunological background" could lead to important differences in the response to cholera infection in the (endemic) Thai population (Suntharasamai et al. 1992; Pitisuttithum et al. 2002). It was thus intended to (i) investigate hypotheses regarding host-pathogen interactions in semi-immune individuals (which would be infeasible in a non-endemic setting), (ii) develop models of infection in an endemic population against which interventions

Table 5.1 Cholera and *Shigella* HCS in Thailand

Pathogen	Cholera (el Tor)	Cholera (O139)	Shigella sonnei	Shigella sonnei
Year of publication	1992	2002	2012	2016
First author	Suntharasamai	Pitisuttithum	Bodhidatta	Pitisuttithum
Country	Thailand	Thailand	Thailand	Thailand
Consent	Test of understanding	Test of understanding	Test of understanding	Test of understanding
Follow-up after study	N/S	N/S	N/S	2 months
Mode of challenge	Oral	Oral	Oral	Oral
Source of challenge strain	Laboratory strain	Laboratory strain	Laboratory strain	Laboratory strain
Diagnosis/treatment initiation	Symptoms plus cultures	Symptoms plus cultures	Symptoms plus cultures	Symptoms plus cultures
Treatment	Antibiotics, IV/oral fluids for diarrhoea	Antibiotics, IV/oral fluids for diarrhoea	Antibiotics, IV/oral fluids for diarrhoea	Antibiotics, IV/oral fluids for diarrhoea
Clinical attack rate	90%	100%	75%	20% (controls)
Severe symptoms	Nil	Yes, Up to 16L diarrhoea	Yes, severe dysentery in 3 (8.3%)	1 (16%) severe dysentery in control
Other burdens	Dysentery, isolation	Dysentery, isolation	Dysentery, isolation	Dysentery, isolation
Reduced severity in endemic population	Yes	Yes	N/S	Unclear; less shedding of vaccine strain
In- or outpatient	Inpatient	Inpatient	Inpatient	Inpatient
Third party risks/mitigation	Released once stool cultures negative	Released once stool cultures negative	Released once stool cultures negative	Released once stool cultures negative
Long-term effects	N/S	N/S	N/S	N/S
Pathogen	Cholera (el Tor)	Cholera (O139)	Shigella sonnei	Shigella sonnei
Year of publication	1992	2001	2012	2016
First author surname	Suntharasamai	Pitisuttithum	Bodhidatta	Pitisuttithum
Country	Thailand	Thailand	Thailand	Thailand
Location (endemic or not)	Bangkok (endemic)	Bangkok (endemic)	Bangkok (endemic)	Bangkok (endemic)
International Collaborators	Nil	USA, WHO	USA	USA
Prior HIC HCS	Yes	Yes	Yes	Yes
Pre-registration	N/S	N/S	N/S	Clinicaltrials.gov

(continued)

Table 5.1 (continued)

Pathogen	Cholera (el Tor)	Cholera (O139)	Shigella sonnei	Shigella sonnei
Year of publication	1992	2001	2012	2016
First author surname	Suntharasamai	Pitisuttithum	Bodhidatta	Pitisuttithum
Country	Thailand	Thailand	Thailand	Thailand
Community engagement	N/S	N/S	N/S	N/S
Ethics/IRB	Local review	Local review	Local review	Local review, US Army IRB
Challenge strain regulation	N/S	N/S	N/S	FDA CGMP (vaccine)
Rationale	Model development	Model development	Model development	Vaccine trial
n	26	35	36	20 (6 controls)
% female	0%	31.43%	36.11%	N/S
Participants' background	N/S	85% labourers, unemployed, students	N/S	N/S

N/S not specified, *IV* Intravenous, *WHO* World Health Organisation, *HLA-B27* a genetic marker of the risk of post-infectious arthritis
'International collaborators' were derived from institutional affiliation of authors only

could be tested, and (iii) respond to local disease burden priorities. The models developed were intended for use in future local cholera vaccine trials, although no subsequent vaccine trials have cited these studies thus far.

The stated rationales for the *Shigella* studies were more detailed. In addition to assessing different response to challenge in a Thai population, the authors noted (i) the large global burden of *Shigella* (including among Thai children) related to acute disease and its long-term complications (post-infectious bowel symptoms, arthritis), (ii) increasing drug resistance of *Shigella*, and (iii) the lack of an animal model (because humans are the only natural host of the pathogen). The model was developed in the 2012 *Shigella* HCS and subsequently used in 2016 to test a vaccine. The studies were reviewed and approved by the local institutional ethics committee, the Ministry of Public Health (Thailand), and, in the case of the *Shigella* studies, the U.S. Army Human Subjects Research Review Board. The vaccine used in the 2016 study was produced according to US FDA cGMP and presumably approved for research use by the Thai FDA. The challenge strains used were not governed by specific Thai regulation. The 2016 trial was the first of the Thai studies to be pre-registered on clinicaltrials.gov.

5.1.2 Recruitment, Participant Selection, Consent, and Payment

One of the four studies reported the background of participants: 43% students, 37% labourers, and 20% unemployed (Pitisuttithum et al. 2002). Participants were recruited from a population in which the diseases are endemic, although not all participants had evidence of high levels of past exposure (for example, in the 2012 *Shigella* study 20% of those screened had antibodies to *Shigella*) (Bodhidatta et al. 2012). Exclusion criteria included (i) in all studies: general co-morbidities and pregnancy (due to potential increased risks), (ii) in the *Shigella* studies: carrying HLA-B27 (due to the associated risk of post-infectious arthritis) (iii) in the vaccine trial: those with abnormal bowel habit (which might reduce the risk of post-infectious irritable bowel syndrome and/or improve the clarity of vaccine efficacy estimates). Cholera severity is known to vary with blood type, thus the cholera studies recruited individuals with a variety of blood types in order to increase generalisability of findings. Participants also underwent psychological screening by investigators to select those who would tolerate inpatient isolation. All Thai HCS involved a written test of comprehension as part of the informed consent process. Payment is not recorded in the publications, but was apparently indexed to local wages for unskilled labour and the period of isolation in the inpatient unit (Personal communications from interview participants).

5.1.3 Burdens (Including Risks to Participants and Third Parties)

If required, symptomatic participants (e.g., those with diarrhoea and/or dysentery) received oral or intravenous rehydration. Among subjects who became symptomatic, there were few severe symptoms, although stool volumes in the cholera studies ranged from around half a litre to up to 16 litres. On average, symptoms were milder than in similar studies in non-endemic North American populations, suggesting that enteric pathogen HCS may at least sometimes be less burdensome in an endemic population.

Post-infectious complications, including irritable bowel syndrome, reactive arthritis, and Guillain-Barre syndrome, have been observed after *Shigella* infection (Thabane et al. 2007; Hannu 2011; Ajene et al. 2013) but they are rare and have not been reported to occur among HCS participants. The probability of such complications can be difficult to predict in the absence of known risk factors. Reactive arthritis has the known risk factor of HLA-B27 (a genetic marker) (Hannu 2011); thus those with this marker were excluded to avoid excessive risks—although even those who do not carry HLA-B27 have a small risk of post-infectious arthritis and other complications. Scientists interviewed for this study were not aware of any long-term complications related to study participation.

The studies were conducted in fully inpatient settings with strict biosafety procedures, decontamination of effluent, and treatment of participants (with proof of cure) to reduce third-party risks. In the *Shigella* trial, among those vaccinated 56% were found to shed the vaccine strain in stool, which could potentially spread to others, although no cases of transmission were reported in a previous Israeli study of this vaccine (Orr et al. 2005).[1]

5.1.4 Summary and Outcomes

The Thai Cholera HCS were noteworthy in the sense that they demonstrated that such studies could be successfully and safely conducted in an endemic LMIC. The model was able to demonstrate differences in symptomatic disease in an endemic population, but it has not yet been used to test novel interventions for cholera. The *Shigella* model was relatively quickly advanced to a vaccine trial, however the latter was not able to provide accurate estimates of vaccine efficacy because the attack rate among controls (given placebo) was 20% (i.e., a failure to replicate the attack rate of 75% in the HCS in which the model was developed (Bodhidatta et al. 2012))—since so few of those who were not vaccinated developed disease, the estimate of vaccine efficacy (i.e., equal to any further reduction in risk of disease post-challenge in vaccines) was not statistically significant. We have been informed that Thai researchers are considering undertaking more HCS in the future (Personal communications from study participants).

5.2 Falciparum Malaria Challenge Studies in Africa

We identified 5 Sub-Saharan African HCS involving falciparum malaria in Tanzania, Kenya, and Gabon (Hodgson et al. 2014; Shekalaghe et al. 2014; Lell et al. 2017; Dejon-Agobe et al. 2018; Jongo et al. 2018). These studies are part of an international research program using cryopreserved *P. falciparum* (Pf) sporozoites from a laboratory strain (NF54) known to be sensitive to chloroquine and produced in increasingly controlled laboratory settings since the 1980s (Chulay et al. 1986). The research program began at the US Walter Reed Army Institute of Research (WRAIR) and is now led by a US private company, Sanaria Inc., in collaboration with several research centres worldwide. This laboratory strain has been tested in multiple studies (in both HICs and LMICs) where it has been used as (i) a human challenge agent (non-attenuated) and (ii) a vaccine candidate (radiation attenuated), administered either by injection (intradermal, intramuscular, or

---

[1] Oral live attenuated vaccines, such as the oral polio vaccine, can immunise others in the community when spread via the stool of those vaccinated.

intravenous) or by mosquito bite (Lyke et al. 2010; Hodgson et al. 2014; Shekalaghe et al. 2014; Gómez-Pérez et al. 2015; Olotu et al. 2018).

The US FDA-approved NF54 challenge strain is produced according to cGMP regulations designed to ensure purity (only the sporozoite form of malaria parasites, only one strain) and asepsis (not containing bacterial or other pathogens). After multiple models and trials in non-endemic-regions, this research group, with international collaborators, has more recently been conducting endemic-region falciparum malaria HCS, resulting in the sub-Saharan African publications reviewed below (See also Table 5.2). One advantage of this model is that it does not require mosquitoes to administer the infection challenge (which is instead injected by needle) thus obviating the need for high biosafety level insectaries (which are rare in LMICs and involve significant costs) and/or the need to import mosquitoes (which might involve risks to the local population, as discussed in Sect. 3.4) (Billingsley et al. 2014); whether malaria HCS by needle is generalisable to wild-type infection by mosquito bite is unknown, but the model uses sporozoites (i.e., the same form of malaria parasite involved in transmission from mosquitoes to humans) in order to investigate the early stages of malaria infection (and, for example, whether these can be prevented or reduced by novel interventions).

5.2.1 Rationale and Review Process

This program of sub-Saharan African HCS research, conducted in conjunction with HIC collaborators, was designed to (i) test hypotheses regarding host-pathogen interactions in malaria-endemic and/or semi-immune populations (that would have been infeasible in HIC HCS) as well as those with genetic traits thought to affect malaria infection (e.g. sickle cell and α-thalassaemia), (ii) develop models of infection in an endemic population against which to test interventions (thus aiming to improve the generalisability of malaria HCS to African populations), (iii) test malaria interventions (e.g. vaccines) against these models, (iv) respond to local disease burden priorities, and (v) further develop local capacity for infectious disease research (at well-established African institutions in all three countries with significant pre-existing research expertise). Four of the five studies were focused on investigating host-pathogen interactions and/or developing Africa-specific infection models as translations of models that had previously been conducted in non-endemic HICs. The fifth study used the locally-tested model to trial a malaria vaccine candidate (GMZ2) against challenge (Dejon-Agobe et al. 2018).

All five studies were pre-registered (with US clinicaltrials.gov and/or the Pan-African Clinical Trials registry). Ethics committees at the African institutions and collaborating HIC institutions reviewed the studies, which were also reviewed by national committees in the African countries. The challenge strain had prior regulatory approval by the US FDA, and was reviewed by Kenyan, Tanzanian, and Gabonese regulators. The studies took place at three research institutes with longstanding African-HIC collaborative international research programs.

Table 5.2 Falciparum malaria HCS in Africa

Pathogen	Malaria (falciparum)	Malaria (falciparum)	Malaria (falciparum)	Malaria (falciparum)	Malaria (falciparum)
Year of publication	2014	2014	2018	2018	2018
First author	Hodgson	Shekalaghe	Jongo	Lell	Dejon-Agobe
Country	Kenya	Tanzania	Tanzania	Gabon	Gabon
Location (endemic or not)	Nairobi (non-endemic)	Bagamoyo (endemic)	Bagamoyo (endemic)	Lambaréné (endemic)	Lambaréné (endemic)
Type of study	Infection model	Infection model	Vaccine trial	Infection model	Vaccine trial
International collaborators	UK, USA	Netherlands, Switzerland, USA	Switzerland, USA	Germany, USA	Germany, Netherlands, USA
Prior HIC HCS	Yes	Yes	Yes	Yes	Yes
Pre-registration	Pan African clinical trial registry	Clinicaltrials.gov	Clinicaltrials.gov	Clinicaltrials.gov	Pan African clinical trial registry
Community engagement	Yes—extensive publication of process	N/S	N/S	N/S	N/S
Ethics/IRB	Local, national, and international review	Local, national, and international review	Local, national, and international review	Local, national, and international review	Local, national, and international review
Challenge strain regulation	Kenyan Pharmacy and Poisons Board, US FDA	Tanzania FDA, US FDA	Tanzania FDA, US FDA	Direction Médicament et de la Pharmacie, US FDA	Direction Médicament et de la Pharmacie, US FDA
Rationale	Model development	Model development	Vaccine trial	Model development	Vaccine trial
n	28	30 (6 controls)	67 (18 controls)	25	50 (10 controls)
% female	39.29%	0%	0%	48%	0%
Recruitment details	54% student, 17% unemployed, 29% other	Students at higher learning institutions	Students at higher learning institutions	N/S	N/S
Pathogen	Malaria (falciparum)	Malaria (falciparum)	Malaria (falciparum)	Malaria (falciparum)	Malaria (falciparum)
Year of publication	2014	2014	2018	2018	2018
First author surname	Hodgson	Shekalaghe	Jongo	Lell	Dejon-Agobe
Country	Kenya	Tanzania	Tanzania	Gabon	Gabon
Consent	Test of understanding, time for consideration	Test of understanding, time for consideration	Test of understanding, time for consideration	Test of understanding, time for consideration	Test of understanding, time for consideration

(continued)

Table 5.2 (continued)

	Malaria (falciparum)	Malaria (falciparum)	Malaria (falciparum)	Malaria (falciparum)	Malaria (falciparum)
Pathogen	Malaria (falciparum)	Malaria (falciparum)	Malaria (falciparum)	Malaria (falciparum)	Malaria (falciparum)
Year of publication	2014	2014	2018	2018	2018
First author surname	Hodgson	Shekalaghe	Jongo	Lell	Dejon-Agobe
Country	Kenya	Tanzania	Tanzania	Gabon	Gabon
Follow-up post-challenge	Exit questionnaire	Up to 168 days for recurrent malaria; 1 infection during follow-up (not challenge strain)	N/S	Up to 3 months for recurrent malaria	8 weeks
Mode of challenge	Intramuscular	Intradermal	Intravenous	Intravenous	Intravenous
Pathogen strain	NF54 Lab. strain	NF54 Lab. strain	NF54 Lab. strain	NF54 Lab. strain	NF54 Lab. strain
Diagnosis	Symptoms + microscopy	Symptoms + microscopy	Symptoms + microscopy	Symptoms + microscopy OR qPCR	Symptoms + microscopy
Treatment	Antimalarial	Antimalarial	Antimalarial	Antimalarial (pre- and post-)	Antimalarial (pre- and post-)
Clinical attack rate	100%	75%	39% (controls)	Europeans (100%), Sickle cell Gabonese (14%), Other Gabonese (67%)	44% (overall)
Severe symptoms n (%)	17 (61%)	1 (3%)	3 (4.5%)—all fever only	Europeans: 3 (60%), Gabonese: 0 (0%)	Nil
Other burdens	Inpatient isolation, monitoring	Inpatient isolation, monitoring	Inpatient isolation, monitoring	Monitoring	Monitoring
Reduced severity in endemic population	N/S	Yes	Yes	Yes	Yes
In- or outpatient	Inpatient	Partly outpatient	Inpatient post-challenge (12 days)	Outpatient	Outpatient

N/S Not Specified, *ICH* International Conference on Harmonisation, *MoH* Ministry of Health, *AEs* Adverse Effects
'International collaborators' were derived from institutional affiliation of authors only

The Kenyan group published ethics committee and regulatory approval times (local Kenyan committee: 6 months; collaborating UK university committee: 2 months; Kenyan Pharmacy and Poisons Board: 3 weeks) (Hodgson et al. 2015). Since the timeliness (as well as thoroughness) of ethical review can itself be ethically important (e.g., because undue delays of beneficial research arguably delay benefits of new interventions for the eventual target population), more standardised reporting of review times across multiple studies might be called for.

5.2.2   Recruitment, Participant Selection, Consent, and Payment

The Kenyan study was conducted in Nairobi, a non-endemic part of the country and the investigators found it difficult to recruit as many semi- and/or highly-immune individuals as they had planned, largely because of a lack of exposure to prior infection in the local population in Nairobi (the research group have since begun conducting HCS in and/or recruiting from more endemic parts of Kenya (Kapulu et al. 2018)). In contrast, the Tanzanian and Gabonese studies took place in endemic areas and were able to recruit semi-immune individuals (and/or those with innate resistance to malaria) more easily. All studies aimed to recruit healthy, non-pregnant, African adults who would share at least some characteristics (e.g., similarities in genetic factors, microbiome, etc.) with high priority target populations for falciparum malaria interventions (e.g., African children). In addition, the studies (to varying degrees) aimed to recruit individuals with acquired immunity (from past malaria infection) in order to study relationships between immunity and response to challenge; the study in Gabon by Lell et al. also recruited a small group of local expatriate Europeans in order to make direct comparisons between European and Gabonese HCS outcomes (Lell et al. 2017).

In order to reduce risks to participants, all five studies excluded those who were pregnant or intending to become pregnant and those with various medically significant comorbidities and/or particular co-infections (e.g., HIV, viral hepatitis). In terms of comorbidities, exclusion criteria included having certain risk factors for cardiac events (based on previous data of cardiac events during malaria HCS (Nieman et al. 2009; van Meer et al. 2014), see Sect. 3.3.5) and psychiatric risk factors (see Sect. 3.3.6.1) (Dejon-Agobe et al. 2018). Investigators in this program had initially planned to exclude individuals with α-thalassaemia trait (because this trait may provide some innate resistance to malaria infection leading to concerns that this could affect HCS results); but they ultimately included these individuals when the high frequency of the trait in the local population became apparent because of the need to avoid (potentially unfair) exclusion of such individuals in future vaccine trials.

Consent in all cases involved a test of understanding. Some studies aimed to recruit relatively well-educated individuals, with a particular focus in the Kenyan study on recruiting medical students, because it was expected that such individuals would be more able to understand the study and thus provide adequately informed consent (Hodgson et al. 2014; Shekalaghe et al. 2014; Hodgson et al. 2015). Although the Kenyan study only conducted information sessions for prospective participants with medical students (from whom the majority of the study population was expected to be recruited), the final study population was mixed (54% students, 17% unemployed, 29% other); the authors concluded that "there was no clear advantage to exclusively targeting medical students and future studies would appeal to students of all disciplines." (Hodgson et al. 2015). Given that non-students and unemployed individuals were also able to pass a test of understanding as part of the consent process, these findings might also support the recruitment of (a larger proportion of) the general adult population in future studies. In Tanzania, all participants were drawn from higher learning institutions, and 100% were male (reflecting lower local rates of higher education among females in general (Kilango et al. 2017)).

The Kenyan group published payment amounts in an article highlighting "lessons learnt" from their first HCS. Participants were paid around $50 USD per overnight stay, as well as smaller amounts for clinic visits and travel costs, amounting to total payments of around $250-$500 USD (Hodgson et al. 2015; Nordling 2018). Payment levels were carefully considered by both local and UK ethics committees, which judged that decided amounts would "neither unduly coerce potential participants nor set a difficult precedent for other research conducted within the programme" (Hodgson et al. 2015; Njue et al. 2018). Payment did, however, lead to short-lived controversy in the local media (Gathura 2018; Kenya Medical Research Institute (KEMRI) 2018). In response to media coverage, the local research institution issued a Statement that included details of the study (including the study rationale, the reasons for inpatient monitoring, the minimisation of risks to participants, and the prior approval by relevant ethical review bodies) and information regarding determinations of the level of payment. The Statement notes that "[T]he participants were compensated for the time they spent at the in-patient facility. The amount compensated was arrived at by considering what they would earn on a daily basis were they engaged in their daily earning activities." (Kenya Medical Research Institute (KEMRI) 2018).

As a comparison, the HCS in Gabon paid participants similar amounts indexed to local wages (personal communication, expert stakeholder), although the Gabonese study was an outpatient design and payment figures have not been published; in any case payment has not led to controversy in that setting. One Kenyan researcher interviewed for this project indicated that, despite the media controversy, local researchers felt that the level of payment was appropriate—although the group did consider a series of smaller payments rather than one large payment at the end of the study:

> [Payment has] been controversial, I think, because of the amount that ended up being given. And it's not so much the daily amount than the lump sum amount – and because it ended up being quite a substantial figure, if you add it up. But if this was being given on a daily basis probably during the study [it wouldn't have seemed] that high … What changed? Might we

have reconsidered given the [controversy in the media]? I think we still feel that rate was fair, in our context. And I don't think there were any plans to lower it, because it wouldn't make sense if you consider the fact that [it was based on national] minimum wage … for casual labour … probably what might ever been considered is … whether to give it as a lump sum or give it periodically instead of building up into a nice packet at the end of the day. I can see participants will want a lump sum, because then they can do something practical. But, we fear that [this large amount] was again being seen … people focus on the lump sum amount more than what the participants need. [Scientist involved in the Kenyan HCS]

5.2.3   Burdens (Including Risks to Participants and Third Parties)

The rate of severe malaria symptoms was generally lower in African individuals (with the exception of the Kenyan study, discussed below), especially those with innate or acquired immunity, than in previous HIC falciparum malaria HCS. In this sense these endemic LMIC malaria HCS presented relatively lower risks to participants than HIC HCS. Arguably, the use of more sensitive diagnostic methods (e.g., PCR) could have reduced risks in terms of symptoms still further, and perhaps helped to support the case for outpatient studies. For example, in some previous HIC malaria HCS, if a participant had symptoms without malaria parasites on blood microscopy, investigators had access to quantitative polymerase chain reaction (qPCR) testing for malaria (a more sensitive test than microscopy) and initiated treatment if this was positive (Sheehy et al. 2013). Timely qPCR was not locally available in these African studies (although samples were tested using such methods at a later date for research purposes). In any case, treatment was effective and no participant experienced treatment failure or recurrent (study-related) malaria during the follow-up period.

A large proportion (61%) of Kenyan participants ultimately developed at least one severe symptom of malaria—a rate similar to European participants, perhaps in part due to the non-endemic setting in Nairobi—but none required hospital care. In contrast, the rate of severe symptoms was much lower among Tanzanian and Gabonese participants (see Table 5.2), potentially because they were drawn from more endemic populations (and thus had greater immunity).

The Gabonese studies took place in an endemic area and predominantly involved outpatient design. Likewise, the Tanzanian studies took place in an endemic area and used a mixed out- and inpatient design. The Kenyan study conducted in Nairobi (a non-endemic area) employed impatient design partly because traffic-related delays could prevent timely access to medical care and partly because investigators were being especially cautious as this was the first HCS performed in Kenya. One member of the research team described decisions related to the design of the study as follows:

I would have been happy to do the Kenyan study as an outpatient setting [but] the traffic was a massive issue … [And especially because this] was a pilot study, [the] first one in the country. [People] said, 'There will be a lot of anxiety about this' … [T]he idea was that we

would need to [have close inpatient monitoring] to provide reassurance [to]the ethical bodies
… but, you know, it was terrifying, actually, when you're travelling to and from the setting,
because the traffic was such that if there was an ambulance coming through, the traffic was
so gridlocked, you couldn't actually physically get an ambulance along any of the roads.
[Scientist involved in the Kenyan HCS]

In terms of risk to third parties, universal treatment upon diagnosis and/or at the
end of the study period and the short duration of challenge infection reduced risks
significantly. Duration of malaria infection affects third-party risk because
gametocytes, the transmissible form of malaria, take 7–15 days to develop (Roberts
et al. 2013)—meaning that risks to third parties increase with time since infection.
The Kenyan study posed the lowest risk of transmission (effectively zero risk), both
because it was an inpatient study and because of a lack of mosquito vectors in
Nairobi. In the 2014 Tanzania study, participants were kept as inpatients for
21 days and any who had not been diagnosed with malaria by this time were
discharged and continued to participate as outpatients until day 28, at which time
they were treated regardless of whether they met the diagnostic threshold. There
was a (perhaps very) low probability of transmission of malaria in such
circumstances (Karl et al. 2011; Roberts et al. 2013) (and this could, in future, be
quantified by qPCR measurement of gametocyte levels during study participation).
The Gabon study was conducted on a largely outpatient basis and likewise may
have involved a low probability of transmission from study participants to the
wider community, in the context of a high local background malaria transmission.

5.2.4 Summary and Outcomes

Among these studies were two significant milestones for LMIC HCS research: the
first HCS in Africa since the 1950s (Allison 1954; Bearcroft 1956; Shekalaghe
et al. 2014), and the first HCS involving vaccine efficacy testing in Africa
(Dejon-Agobe et al. 2018). In terms of scientific outcomes, the research program
has clarified the protective effects of acquired immunity (which led to a delay to
onset of parasitaemia), sickle cell trait (which was shown to be associated with
lower levels of symptomatic malaria), and α-thalassemia (which was less protective
than anticipated), which might support the recruitment of such groups to future
vaccine efficacy trials.

Authors of the Kenyan study noted that one participant had (qPCR positive,
microscopy negative) asymptomatic parasitaemia (which was successfully treated)
at the end of the study, and that future (endemic region) studies of longer duration
could investigate the transmissibility of such asymptomatic infections in partially
immune individuals (see also Vallejo et al. 2016). However, they also note that this
would entail ethical trade-offs involving difficult choices between (i) long periods
of inpatient isolation for participants or (ii) potential risks of transmission to third
parties where an outpatient model is used and there are local mosquito vectors.

The Kenyan group published an extensive account of "lessons learnt" during and after this HCS. Such lessons included (i) the importance of prospective multi-stakeholder engagement and listening to the concerns of the local community, (ii) the need for extensive information sessions for participants to ensure that they were able to understand the study, (iii) the need to include local sub-populations (e.g. those with haemoglobinopathies) so that they would not be unfairly excluded from the benefits of HCS research, (iv) the need for longer duration studies of naturally-acquired immunity (since semi-immune participants took longer to develop an infection after challenge) and models of transmissibility, and (v) the importance of support from other centres with prior experience conducting malaria HCS in helping to ensure the safety and efficiency of the Kenyan study (Hodgson et al. 2015), Researchers at the same institution have also subsequently published social science work related to malaria HCS (Njue et al. 2018). The publication of such insights from experience with HCS, as well as the integration of biological and social science work related to HCS, could be considered a model of best practice in terms of engagement and the sharing of ethically relevant practical details that might inform future HCS designs.

5.3 Vivax Malaria Challenge Studies in Colombia

The 4 published vivax malaria HCS in Colombia, beginning in 2009 (the earliest malaria HCS in an endemic country that we identified apart from the historical cases discussed above—See Sect. 2.5) were conducted by a well-established research group in Cali (a non-endemic city with endemic areas of transmission relatively nearby–within a few hours' drive). The local institution has been involved in malaria research for many decades, including, for example, maintaining a longstanding malaria vector mosquito insectary, which provided the mosquitoes for these studies. The HCS formed part of a local research program, one of the goals of which is the development of vivax malaria interventions, especially vaccines.

Since there is no available laboratory strain of vivax malaria, the endemic country setting of the institution enabled researchers to obtain wild-type malaria parasites from consenting patient donors infected in nearby endemic parts of the country (immediately prior to these patients receiving treatment), transport these parasites to Cali, infect insectary-reared mosquitoes (after careful screening for other blood-borne infections), and challenge HCS participants.

5.3.1 Rationale and Review Process

This research program, led by local researchers (in some cases in collaboration with international scientists), was designed to (i) investigate hypotheses regarding host-pathogen interactions, including in semi-immune individuals (i.e. study designs that are only feasible in a study centre in/near an endemic setting), (ii) develop a model of infection against which to test interventions, (iii) test novel interventions (e.g. a vivax malaria vaccine), (iv) respond to local disease burden and maximise generalisability by using local wild-type parasites, and (v) ensure minimisation of third-party risks by conducting HCS in a non-endemic area of Colombia with no local vector mosquitoes (only laboratory mosquitoes were used under strict biosafety precautions). The research program began by testing a new vivax HCS model in malaria naïve individuals in 2009 (Herrera et al. 2009). This model was later refined in 2011 (Herrera et al. 2011), tested in semi-immune individuals in 2014 (Arévalo-Herrera et al. 2014), and ultimately used to test a vaccine in 2016 (Arévalo-Herrera et al. 2016). From 2014, the studies were pre-registered on clinicaltrials.gov.

The studies were reviewed and approved by local institutional ethics committees and, where there was significant US collaboration, by US committees at collaborating institutions. The first study was also reviewed by WHO. Challenge organisms are not governed by specific Colombian regulations; interviewed scientists with knowledge of the studies described significant efforts that were made to ensure that the laboratory environment (including insectary-mosquitoes) and the challenge material were as close to FDA-style GMP as possible (including extensive screening of donor blood), noting that full compliance with stringent GMP requirements (such as those used for the NF54 lab strain in the African studies above) would not be possible in the absence of a laboratory strain of vivax.

One regulatory issue that affected the group was applying for insurance (e.g., for research-related harm to participants). Local Colombian insurers (backed by international, usually North American, reinsurers) were initially reluctant to cover the research, which created a delay of approximately two years, as described by a member of the research team:

> [Insurers didn't] want to provide the insurance … they are thinking that maybe … we are not designing the [study] protocol [well], or that the volunteers are at high risk of [problems related to] safety. So, to convince them, it was very difficult for us … For the first clinical trial, the phase one … it took us like two years [to get] that insurance … [Since we got it, and they know us] we [have been able to] renew our insurance without problems. [Myriam Arévalo-Herrera, scientist, Colombia]

The researchers have never had to make a claim against this insurance (since no lasting harms have occurred among volunteers), however this experience highlighted an additional practical issue that may sometimes be more difficult in LMICs than HICs (Table 5.3).

5.3.2 Recruitment, Participant Selection, Consent, and Payment

The studies recruited healthy adults from the general population with a focus on malaria-naïve and/or semi-immune volunteers depending on the research question. The authors note the ethically relevant point that, because new interventions for

Table 5.3 Vivax malaria challenge studies in Colombia

Pathogen	Malaria (vivax)	Malaria (vivax)	Malaria (vivax)	Malaria (vivax)
Year of publication	2009	2011	2014	2016
First author surname	Herrera	Herrera	Arévalo-Herrera	Arévalo-Herrera
Country	Colombia	Colombia	Colombia	Colombia
Location (endemic or not)	Cali (no vectors)	Cali (no vectors)	Cali (no vectors)	Cali (no vectors)
Pre-registration	N/S	N/S	Clinicaltrials.gov	Clinicaltrials.gov
Type of study	Infection model	Infection model	Infection model, test of acquired immunity	Vaccine trial
International collaborators	Brazil, USA, WHO	Brazil, USA	Brazil	USA
Prior HIC HCS	No	No	No	No
Community engagement	N/S	N/S	N/S	N/S
Ethics/IRB	Local and international review	Local and international review	Local review	Local review
Challenge strain regulation	N/A	N/A	N/A	N/A
Rationale	Model development	Model development	Model development	Vaccine trial
n	18	22 (5 controls)	16	28
% female	50%	52.9%	37.5%	64%
Recruitment	Malaria-naïve	Malaria-naïve	Malaria-naïve and semi-immune	Malaria-naïve
Pathogen	Malaria (vivax)	Malaria (vivax)	Malaria (vivax)	Malaria (vivax)
Year of publication	2009	2011	2014	2016
First author surname	Herrera	Herrera	Arévalo-Herrera	Arévalo-Herrera
Country	Colombia	Colombia	Colombia	Colombia
Consent	Multiple information sessions, standard consent	Multiple information sessions, standard consent	Test of understanding	Test of understanding
Follow-up post-challenge	18 months	1 year	3 months	2 months

(continued)

Table 5.3 (continued)

Pathogen	Malaria (vivax)	Malaria (vivax)	Malaria (vivax)	Malaria (vivax)
Year of publication	2009	2011	2014	2016
First author surname	Herrera	Herrera	Arévalo-Herrera	Arévalo-Herrera
Country	Colombia	Colombia	Colombia	Colombia
Mode of challenge	Mosquito + donor blood	Mosquito + donor blood	Mosquito + donor blood	Mosquito + donor blood
Pathogen strain	Wild type	Wild type	Wild type	Wild type
Diagnosis/treatment initiation	Microscopy	Microscopy + retrospective qPCR	Microscopy + qPCR	Microscopy + qPCR
Treatment	Antimalarial plus primaquine	Antimalarial plus primaquine	Antimalarial plus primaquine	Antimalarial plus primaquine
Clinical attack rate	94%	100%	100% in malaria naïve, 33% in semi-immune	7% in vaccines, 71% in infectivity controls (duffy negative)
Mild-moderate symptoms	Symptoms of malaria	Symptoms of malaria	Symptoms of malaria	N/S
Severe symptoms	Nil	1 (anxiety crisis)	Some with severe malaria symptoms	N/S
Reduced severity in endemic population	N/A	N/S	Yes	N/A
In- or outpatient	Outpatient	Outpatient	Outpatient (except semi-immune individuals)	Outpatient
Long-term effects	N/S	N/S	N/S	N/S

N/S Not Specified, *ICH* International Conference on Harmonisation, *FDA* US Food and Drug Administration
'International collaborators' were derived from institutional affiliation of authors only

malaria (e.g., vaccines) will potentially be used in semi-immune individuals, such individuals should arguably be included in at least some HCS research (an argument also made in support of the recruitment practices of African malaria HCS discussed above). Exclusion criteria were predominantly designed to reduce risks (e.g., G6PD deficiency—a risk factor for adverse effects with primaquine treatment of vivax, HIV and other major co-infections, and pregnancy).

The consent process allowed time for consideration across multiple sessions, and prospective participants were encouraged to discuss their participation with family members. A written test of understanding was used from 2011 onwards. The authors emphasise that participants could withdraw from the studies at any time. This did sometimes occur before challenge, but there were no withdrawals after challenge. Basic literacy was required, because individuals who could not read the study material and consent form were excluded. However, unlike some other challenge studies (see African HCS above) where medical students were considered ideal candidates, the research group tended to avoid preferential recruitment from the healthcare sector after an unfortunate episode in an earlier study where a participant, who was elsewhere employed as a paramedic, was suspected to have self-treated with anti-malarial medication after challenge (Herrera et al. 2009).

Participants in the Colombian HCS were reimbursed for their costs related to participation but, as per Colombian norms, did not receive any further payment. Though they did not financially incentivise participation, the research group reports no difficulty recruiting volunteers. One member of the research team attributed recruitment success to local experiences of clinical malaria and altruism among the local population:

> [T]he difference between here, and [the USA] is that [participants in the USA are] doing that for money and here they are doing it because they are convinced, they are altruists. They have seen people suffering from malaria and they want to contribute to solve the problem. It's a significant difference between a volunteer in the States and a volunteer in Colombia. They know we cannot provide any payment, but do it because they are convinced, not because they need money. [Sócrates Herrera, scientist, Colombia]

5.3.3 Burdens (Including Risks to Participants and Third Parties)

Since the Colombian HCS used wild-type vivax parasites from infected human donors, the authors document an extensive review of possible risks to participants (and/or uncertainties) related to the challenge infection. In consultation with independent experts, they reached a consensus that there are no known cases of non-malaria pathogens being transmitted by *Anopheles* mosquitoes, but the consent process included a discussion of uncertainty related to "exposure to potential unknown pathogens" (Herrera et al. 2009). Blood/parasite donors were tested for known pathogens (as per usual blood bank screening) and excluded if any were detected. Mosquitoes that had fed on blood with co-infections (e.g., falciparum malaria with vivax, or viral hepatitis) were discarded. It is likely that multiple vivax strains were transmitted by challenge; although, to our knowledge, this was not tested. One advantage of not testing/screening for this is better generalisability to wild-type infection. The presence/transmission of multiple strains, furthermore, does not necessarily increase the risk of symptoms in participants challenged—and local rates of antimalarial resistance are low (in particular, there is no known resistance to cure of the dormant form of vivax). Physical severe adverse effects were rare, but one patient was admitted to hospital overnight (which meets criteria for a severe adverse effect) with an anxiety crisis (see Sect. 3.3.6.1).

With respect to third-party risk, the local city (Cali) is at relatively high altitude and has no local vector mosquitoes, which minimises the risk of transmission. The mosquitoes used for challenge were laboratory-reared, and any escaping the insectary would face a climate inimical to their survival. The study was conducted on an outpatient basis, and the authors requested that participants avoid travel to areas with vectors while infected, in order to minimise transmission to others. Parasitaemia cleared rapidly with treatment, suggesting that post-treatment transmission risks would have been minimal (if participants later had contact with vectors). Follow-up after the studies was of long duration (up to 18 months reported). Three months after the 2014 study, one participant developed vivax malaria and was treated appropriately. Rather than reactivation of the challenge infection, this was presumed to be a new case of malaria resulting from recent travel to an endemic area—but further testing (e.g., genotyping) to confirm this was not undertaken (Arévalo-Herrera et al. 2014). Local researchers interviewed for this project indicated that there has never been a case of vivax relapse judged to be caused by challenge infection in Colombia.

5.3.4   Summary and Outcomes

The Colombian vivax studies represent a particularly longstanding locally-initiated LMIC HCS program (with support from international collaborators) that, in response to local disease burden, has successfully moved from model development to vaccine testing. The radiation-attenuated vaccine tested by HCS in 2016 showed protective efficacy of 42%, although it required a long and relatively burdensome schedule of immunisation by mosquito bite. At the end of 2018, no field trial had yet cited the 2016 HCS by this group, so it is not yet possible to compare field trial efficacy with that observed in HCS.

Particularly from 2014 onwards, the group has performed a number of secondary analyses on samples collected during HCS, thus maximising the scientific yield per challenge. For example, the researchers used samples from their 2014 study in a later analysis that aimed to quantify any differences of the risk of transmission (by measuring gametocytes, the form of malaria transmitted from humans to mosquitoes) from individuals infected by challenge as opposed to those diagnosed with naturally-acquired infection in the community (Vallejo et al. 2016), which has implications for potential third-party risks of their study design. In contrast to some older data derived from malariotherapy, the study failed to transmit malaria to mosquitoes by feeding them on subjects recently infected by challenge; however, participants from the naturally infected groups were able to transmit malaria to mosquitoes. Since the authors failed to show infectivity of mosquitoes fed on HCS participants, there would have been a very low risk (if any) of onward transmission of the challenge infections during the study period, even if the participants were to leave the study location and travel to an endemic area while infected. Such data may facilitate third-party risk estimates of future vivax malaria HCS involving the infection model developed by this research team.

5.4   Summary of Case Studies

5.4.1   *Rationale and Review Process*

We identified 13 published HCS conducted in LMICs from 1992 to 2108. The number and frequency of LMIC HCS is increasing: 11 of the 13 studies were conducted in the last ten years, and more LMIC HCS are currently being considered and/or conducted. Yet these are still vastly outnumbered by HIC HCS, suggesting that LMIC HCS has been a neglected area of research, especially relative to local disease burden. Each of the 13 studies involved a pathogen endemic to the LMIC in which it was conducted, although in 5 publications the HCS was conducted in a non-endemic area within the country. Common reasons for conducting these LMIC HCS were (i) to improve understanding of host-pathogen interactions in an endemic population, (ii) to develop models of infection (for later HCS vaccine trials), (iii) to test vaccines (in 3 studies), and/or (iv) to improve local capacity for infectious disease research.

All of the studies took place within well-established research institutions that had existing collaborative arrangements with HIC institutions with HCS research experience. With the exception of the Colombia vivax program, most LMIC HCS programs to date (9 of 13 studies) have begun by replicating prior HIC HCS in the local LMIC population. Although the LMIC institutions had experience conducting other types of research, significant capacity building was required in order to conduct HCS, including the capacity of local ethics committees to review such research.

Like most HICs, the LMICs in which HCS have been conducted do not have specific regulations governing challenge strains. Regulatory bodies in Sub-Saharan Africa did review the malaria challenge strain used, aided by previous FDA review and approval. In Thailand and Colombia, the local institutions (and, for the Thai studies, the collaborating US institution(s)) were responsible for the quality and safety of the challenge strains used.

5.4.2 Recruitment, Participant Selection, Consent, and Payment

All LMIC HCS recruited healthy adult volunteers from the local population (and, for one study in Gabon, a sub-population of European expatriates for comparison (Lell et al. 2017)). Depending on the research question, some studies preferentially aimed to include those with acquired immunity from past infection and/or innate resistance to the disease under study—and recruiting from such groups was a notable advantage of conducting the studies in endemic LMICs. Exclusion criteria were designed to reduce risks to participants, including reducing the probability of lasting harm.

In all studies, consent processes involved multiple sessions and/or a formal test of understanding, suggesting a high standard of informed consent. Many studies aimed to recruit students and/or relatively well-educated individuals in order to improve the quality of understanding; however, recent social science work in Kenya has challenged this assumption, suggesting that less educated individuals may be able to provide adequate informed consent (Njue et al. 2018). This is particularly important if/when there are scientific reasons to recruit from highly endemic rural areas in which the average education level may be lower than in (less endemic, or non-endemic) large cities. There are two further issues with recruiting tertiary-educated individuals: firstly, this may lead to the relative exclusion of women, in countries in which women are less likely to receive tertiary education—and this could lead to the results of HCS being less generalisable to women; secondly, students might in some cases feel pressured to agree to participate (e.g., where the HCS is being conducted by researchers from the university/faculty in which the students are studying), which warrants consideration in future HCS designs (whether in HIC or LMIC settings) (Bonham and Moreno 2008).

These LMIC studies, like HIC HCS, involved significant burdens for participants (see below). Payment was usually indexed to burden and to local wages for unskilled labour—with the exception of Colombia, where no payment was offered apart from reimbursement for financial costs incurred by participants.

5.4.3 Burdens (Including Risks to Participants and Third Parties)

LMIC HCS participation involved a range of types and levels of burdens—including being exposed to risk and experiencing symptoms of infection; monitoring, bodily examinations, and blood draws by study staff; time away from normal activities including, in some cases, long periods of inpatient isolation; and so on.

Though generally uncommon, severe physical symptoms did occur—and they were more likely in participants from less endemic populations (e.g., Nairobi) and those without innate resistance or acquired immunity to the pathogen used. One participant required treatment for psychiatric symptoms. Based on published data and interviews with relevant stakeholders, no cases of lasting harm related to LMIC HCS were identified. In some cases the burdens of symptomatic infection could arguably have be further reduced by earlier diagnosis and treatment (e.g., through the use of PCR as opposed to microscopy diagnosis of malaria), particularly where this would not undermine the scientific value of the study.

In terms of risks to third parties, since 5 of the 13 studies were conducted with vector-borne pathogens (falciparum and vivax malaria) in non-endemic areas of LMICs where there are no local vectors, and a further 4 studies were conducted under conditions of strict inpatient isolation (i.e., a total of 9 studies entailed zero or near-zero risks of transmission), only 4 studies posed potential risks of transmission to third parties. These 4 HCS involving malaria in endemic areas of Sub-Saharan Africa, however, would have had low potential for third-party risk due to the short duration of infection and the high local prevalence of, and immunity to, malaria. Some stakeholders felt that such small risks were acceptable, whereas others suggested that they should be reduced still further (see Sect. 3.4).

References

Ajene, A.N., C.L.F. Walker, and R.E. Black. 2013. Enteric pathogens and reactive arthritis: A systematic review of Campylobacter, Salmonella and Shigella-associated reactive arthritis. *Journal of Health, Population, and Nutrition* 31 (3): 299.

Allison, A.C. 1954. Protection afforded by sickle-cell trait against subtertian malarial infection. *British Medical Journal* 1 (4857): 290.

Arévalo-Herrera, M., D.A. Forero-Peña, K. Rubiano, J. Gómez-Hincapie, N.L. Martínez, M. Lopez-Perez, A. Castellanos, N. Céspedes, R. Palacios, and J.M. Oñate. 2014. Plasmodium vivax sporozoite challenge in malaria-naive and semi-immune Colombian volunteers. *PLoS ONE* 9 (6): e99754.

Arévalo-Herrera, M., J.M. Vásquez-Jiménez, M. Lopez-Perez, A.F. Vallejo, A.B. Amado-Garavito, N. Céspedes, A. Castellanos, K. Molina, J. Trejos, and J. Oñate. 2016. Protective efficacy of Plasmodium vivax radiation-attenuated sporozoites in Colombian volunteers: A randomized controlled trial. *PLoS Neglected Tropical Diseases* 10 (10): e0005070.

Baay, M.F.D., T.L. Richie, P. Neels, M. Cavaleri, R. Chilengi, D. Diemert, S.L. Hoffman, R. Johnson, B.D. Kirkpatrick, and I. Knezevic. 2018. Human challenge trials in vaccine development, Rockville, MD, USA, September 28–30, 2017. *Biologicals*.

Bearcroft, W.G.C. 1956. Zika virus infection experimentally induced in a human volunteer. *Transactions of the Royal Society of Tropical Medicine and Hygiene* 50 (5).

Billingsley, P., B.K.L. Sim, E. Bijker, M. Roestenberg, K. Lyke, M. Laurens, B. Mordmueller, P. Gomez, S. Shekalaghe, and S. Hodgson. 2014. Controlled human malaria infections using aseptic, purified cryopreserved Plasmodium falciparum sporozoites administered by needle and syringe. *Malaria Journal* 13 (1): P12.

Bodhidatta, L., P. Pitisuttithum, S. Chamnanchanant, K.T. Chang, D. Islam, V. Bussaratid, M.M. Venkatesan, T.L. Hale, and C.J. Mason. 2012. Establishment of a Shigella sonnei human challenge model in Thailand. *Vaccine* 30 (49): 7040–7045.

Bonham, V.H., and J.D. Moreno. 2008. Research with captive populations: Prisoners, students, and soldiers.

Chulay, J.D., I. Schneider, T.M. Cosgriff, S.L. Hoffman, W.R. Ballou, I.A. Quakyi, R. Carter, J.H. Trosper, and W.T. Hockmeyer. 1986. Malaria transmitted to humans by mosquitoes infected from cultured Plasmodium falciparum. *The American Journal of Tropical Medicine and Hygiene* 35 (1): 66–68.

Dejon-Agobe, J.C., U. Ateba-Ngoa, A. Lalremruata, A. Homoet, J. Engelhorn, O. Paterne Nouatin, J.R. Edoa, J.F. Fernandes, M. Esen, and Y.D. Mouwenda. 2018. Controlled human malaria infection of healthy lifelong malaria-exposed adults to assess safety, immunogenicity and efficacy of the asexual blood stage malaria vaccine candidate GMZ2. *Clinical Infectious Diseases.*

Evers, D.L., C.B. Fowler, J.T. Mason, and R.K. Mimnall. 2015. Deliberate microbial infection research reveals limitations to current safety protections of healthy human subjects. *Science and Engineering Ethics* 21 (4): 1049–1064.

Gathura, G. 2018. Want cash? Volunteer for a dose of malaria parasite, says Kemri amid ethical queries. *The Standard.* Kenya, Standard Group PLC.

Gómez-Pérez, G.P., A. Legarda, J. Muñoz, B.K.L. Sim, M.R. Ballester, C. Dobaño, G. Moncunill, J.J. Campo, P. Cisteró, and A. Jimenez. 2015. Controlled human malaria infection by intramuscular and direct venous inoculation of cryopreserved Plasmodium falciparum sporozoites in malaria-naive volunteers: Effect of injection volume and dose on infectivity rates. *Malaria Journal* 14 (1): 306.

Hannu, T. 2011. Reactive arthritis. *Best Practice and Research Clinical Rheumatology* 25 (3): 347–357.

Herrera, S., O. Fernández, M.R. Manzano, B. Murrain, J. Vergara, P. Blanco, R. Palacios, J.D. Vélez, J.E. Epstein, and M. Chen-Mok. 2009. Successful sporozoite challenge model in human volunteers with Plasmodium vivax strain derived from human donors. *The American Journal of Tropical Medicine and Hygiene* 81 (5): 740–746.

Herrera, S., Y. Solarte, A. Jordán-Villegas, J.F. Echavarría, L. Rocha, R. Palacios, Ó. Ramírez, J.D. Vélez, J.E. Epstein, and T.L. Richie. 2011. Consistent safety and infectivity in sporozoite challenge model of Plasmodium vivax in malaria-naive human volunteers. *The American Journal of Tropical Medicine and Hygiene* 84 (Suppl 2): 4–11.

Hodgson, S.H., E. Juma, A. Salim, C. Magiri, D. Kimani, D. Njenga, A. Muia, A.O. Cole, C. Ogwang, and K. Awuondo. 2014. Evaluating controlled human malaria infection in Kenyan adults with varying degrees of prior exposure to Plasmodium falciparum using sporozoites administered by intramuscular injection. *Frontiers in Microbiology* 5: 686.

Hodgson, S.H., E. Juma, A. Salim, C. Magiri, D. Njenga, S. Molyneux, P. Njuguna, K. Awuondo, B. Lowe, and P.F. Billingsley. 2015. Lessons learnt from the first controlled human malaria infection study conducted in Nairobi, Kenya. *Malaria Journal* 14 (1): 182.

Ioannidis, J.P.A. 2015. How to make more published research true. *Revista Cubana de Información en Ciencias de la Salud (ACIMED)* 26 (2): 187–200.

Jongo, S.A., S.A. Shekalaghe, L.W.P. Church, A.J. Ruben, T. Schindler, I. Zenklusen, T. Rutishauser, J. Rothen, A. Tumbo, and C. Mkindi. 2018. Safety, immunogenicity, and protective efficacy against controlled human malaria infection of Plasmodium falciparum sporozoite vaccine in Tanzanian adults. *The American Journal of Tropical Medicine and Hygiene* 99 (2): 338–349.

Kalil, J.A., S.A. Halperin, and J.M. Langley. 2012. Human challenge studies: A review of adequacy of reporting methods and results. *Future Microbiology* 7 (4): 481–495.

Kapulu, M.C., P. Njuguna, M. Hamaluba, and C.-S.S. Team. 2018. Controlled human malaria infection in semi-immune Kenyan adults (CHMI-SIKA): A study protocol to investigate in vivo Plasmodium falciparum malaria parasite growth in the context of pre-existing immunity. *Wellcome Open Research* 3.

Karl, S., D. Gurarie, P.A. Zimmerman, C.H. King, T.G.S. Pierre, and T.M.E. Davis. 2011. A submicroscopic gametocyte reservoir can sustain malaria transmission. *PLoS ONE* 6 (6): e20805.

Kenya Medical Research Institute (KEMRI). 2018. Response to an article carried in The Standard. Nairobi, Kenya, KEMRI.

Kilango, N.C., Y.H. Qin, W.P. Nyoni, and R.A. Senguo. 2017. Interventions that increase enrolment of women in higher education: The University of Dares Salaam, Tanzania. *Journal of Education and Practice* 8 (13): 21–27.

Lell, B., B. Mordmüller, J.-C.D. Agobe, J. Honkpehedji, J. Zinsou, J.B. Mengue, M.M. Loembe, A.A. Adegnika, J. Held, and A. Lalremruata. 2017. Impact of sickle cell trait and naturally acquired immunity on uncomplicated malaria after controlled human malaria infection in adults in Gabon.

Lyke, K.E., M. Laurens, M. Adams, P.F. Billingsley, A. Richman, M. Loyevsky, S. Chakravarty, C.V. Plowe, B.K.L. Sim, and R. Edelman. 2010. Plasmodium falciparum malaria challenge by the bite of aseptic Anopheles stephensi mosquitoes: Results of a randomized infectivity trial. *PLoS ONE* 5 (10): e13490.

Nieman, A.-E., Q. de Mast, M. Roestenberg, J. Wiersma, G. Pop, A. Stalenhoef, P. Druilhe, R. Sauerwein, and A. van der Ven. 2009. Cardiac complication after experimental human malaria infection: A case report. *Malaria Journal* 8 (1): 277.

Njue, M., P. Njuguna, M.C. Kapulu, G. Sanga, P. Bejon, V. Marsh, S. Molyneux, and D. Kamuya. 2018. Ethical considerations in controlled human malaria infection studies in low resource settings: Experiences and perceptions of study participants in a malaria challenge study in Kenya. *Wellcome Open Research* 3.

Nordling, L. 2018. The ethical quandary of human infection studies. https://undark.org/article/ethical-quandry-human-infection/#comments. Accessed 16 Mar 2019.

Olotu, A., V. Urbano, A. Hamad, M. Eka, M. Chemba, E. Nyakarungu, J. Raso, E. Eburi, D.O. Mandumbi, and D. Hergott. 2018. Advancing global health through development and clinical trials partnerships: A randomized, placebo-controlled, double-blind assessment of safety, tolerability, and immunogenicity of PfSPZ vaccine for malaria in healthy equatoguinean men. *The American Journal of Tropical Medicine and Hygiene* 98 (1): 308–318.

Orr, N., D.E. Katz, J. Atsmon, P. Radu, M. Yavzori, T. Halperin, T. Sela, R. Kayouf, Z. Klein, and R. Ambar. 2005. Community-based safety, immunogenicity, and transmissibility study of the Shigella sonnei WRSS1 vaccine in Israeli volunteers. *Infection and Immunity* 73 (12): 8027–8032.

Pitisuttithum, P., M.B. Cohen, B. Phonrat, U. Suthisarnsuntorn, V. Bussaratid, V. Desakorn, W. Phumratanaprapin, P. Singhasivanon, S. Looareesuwan, and G.M. Schiff. 2002. A human volunteer challenge model using frozen bacteria of the new epidemic serotype, V. cholerae O139 in Thai volunteers. *Vaccine* 20 (5–6): 920–925.

Pitisuttithum, P., D. Islam, S. Chamnanchanunt, N. Ruamsap, P. Khantapura, J. Kaewkungwal, C. Kittitrakul, V. Luvira, J. Dhitavat, and M.M. Venkatesan. 2016. Clinical trial of an oral live Shigella sonnei vaccine candidate, WRSS1, in Thai adults. *Clinical and Vaccine Immunology* 23 (7): 564–575.

Roberts, C.H., M. Armstrong, E. Zatyka, S. Boadi, S. Warren, P.L. Chiodini, C.J. Sutherland, and T. Doherty. 2013. Gametocyte carriage in Plasmodium falciparum-infected travellers. *Malaria Journal* 12 (1): 31.

Sheehy, S.H., A.D. Douglas, and S.J. Draper. 2013. Challenges of assessing the clinical efficacy of asexual blood-stage Plasmodium falciparum malaria vaccines. *Human Vaccines and Immunotherapeutics* 9 (9): 1831–1840.

Shekalaghe, S., M. Rutaihwa, P.F. Billingsley, M. Chemba, C.A. Daubenberger, E.R. James, M. Mpina, O.A. Juma, T. Schindler, and E. Huber. 2014. Controlled human malaria infection of Tanzanians by intradermal injection of aseptic, purified, cryopreserved Plasmodium falciparum sporozoites. *The American Journal of Tropical Medicine and Hygiene* 91 (3): 471–480.

Suntharasamai, P., S. Migasena, U. Vongsthongsri, W. Supanaranond, P. Pitisuttitham, L. Supeeranan, A. Chantra, and S. Naksrisook. 1992. Clinical and bacteriological studies of El Tor cholera after ingestion of known inocula in Thai volunteers. *Vaccine* 10 (8): 502–505.

Thabane, M., D.T. Kottachchi, and J.K. Marshall. 2007. Systematic review and meta-analysis: The incidence and prognosis of post-infectious irritable bowel syndrome. *Alimentary Pharmacology and Therapeutics* 26 (4): 535–544.

Vallejo, A.F., J. García, A.B. Amado-Garavito, M. Arévalo-Herrera, and S. Herrera. 2016. Plasmodium vivax gametocyte infectivity in sub-microscopic infections. *Malaria Journal* 15 (1): 48.

van Meer, M.P.A., G.J.H. Bastiaens, M. Boulaksil, Q. de Mast, A. Gunasekera, S.L. Hoffman, G. Pop, A.J.A.M. van der Ven, and R.W. Sauerwein. 2014. Idiopathic acute myocarditis during treatment for controlled human malaria infection: A case report. *Malaria Journal* 13 (1): 38.

Chapter 6
Conclusions

6.1 Lessons Learned to Date

Perhaps the most significant lesson learned in LMIC HCS to date is that HCS can be conducted safely and to a high scientific standard in LMICs where pathogens such as malaria, cholera, and *Shigella* are primarily endemic. As one stakeholder in Africa noted:

> I think [an important] lesson people have learned is that [although] a number of the partners from the [global] North were very skeptical that this could be done in Africa, now it has been proven that it can be done in endemic countries and it can work well. It's a good way to see that possibly we can start having the whole development process start fairly early in the region and possibly moving this forward. And I think, for me, the feeling is that, we need to sustain doing this so the competency further develops and then we move to the next level and also start getting involved in the entire process of vaccine and even drug development, using the same platform. [Scientist, Africa]

Conducting such research in endemic settings enables HCS that address basic science research questions that would be difficult or impossible to address in non-endemic HICs (e.g., the pathogenesis of infection in partially immune individuals) and the testing of new interventions in a study population from which the results may be more generalisable to the eventual target population for an intervention, with 3 vaccine trial HCS already conducted in LMICs. Furthermore, preparing LMIC institutions to conduct HCS (often in collaboration with HIC institutions) can help to build capacity for infectious disease research and the ethical review of complex HCS designs. As more LMIC institutions are able to conduct HCS for pathogens that are primarily endemic in LMICs, it may arguably become appropriate to prioritise such HCS in LMICs over those in HICs—at least where doing so would improve the generalisability of results and thus facilitate realisation of the ultimate goal of reducing infectious disease burdens in endemic settings.

© The Author(s) 2021
E. Jamrozik and M. J. Selgelid, *Human Challenge Studies in Endemic Settings*,
SpringerBriefs in Ethics, https://doi.org/10.1007/978-3-030-41480-1_6

Many of the unresolved ethical and regulatory issues are common to both LMICs and HICs (see Sect. 6.3), but the particular implications thereof may vary at a local or national level. For example, the optimum models of ethical review of HCS and regulatory review of challenge strains are yet to be determined and may need to be adapted to different settings. However, there are generally much greater funding and capacity constraints for scientific institutions, regulatory bodies and ethics committees in LMICs as compared with HICs, and these should be addressed to assure the quality of HCS conducted in such settings. The success of the studies above has been partly attributed to significant capacity building efforts (Hodgson et al. 2015).

In terms of the burdens of research, HCS in endemic populations with innate or acquired immunity to the challenge pathogen often entail (on average) less risks for participants than those in non-endemic populations with low prevalence (or absence) of such protective factors. Some other burdens of participation were increased in certain settings: for example, relatively long periods of inpatient isolation have been used in endemic settings because of weak local infrastructure that might mean that risks to participants and/or third parties would be unacceptably high with outpatient designs. On the other hand, outpatient HCS were successfully conducted in Colombia (in a non-endemic area) as well as in Tanzania and Gabon (in endemic areas). While the latter involved a (perhaps very low) probability of transmission of malaria to third parties, it is contentious whether such small increases in high (endemic) background risk are of public health importance and/or ethically problematic.

Involving social scientists and community engagement workers in LMIC HCS has been a successful strategy in terms of informing and learning from volunteers and the local community. Such work has generated rich additional data regarding controversial questions (such as whether to recruit by education level and how individuals respond to payment for HCS participation) that will help to inform future HCS design (Njue et al. 2014, 2018). Further (ethical and scientific) improvement of LMIC HCS is particularly important given that such studies constitute unique opportunities for improving scientific knowledge regarding, and developing new interventions to reduce the burden of, neglected diseases primarily endemic in LMICs. Researchers, ethics review committees, and regulators should arguably continue to pay careful attention to HCS design in particular contexts since the success of LMIC HCS partly depends on community acceptance of such research designs, which in turn depends on an acceptable level of burdens for participants (and third parties) and a thorough and transparent review process.

6.2 Points of Consensus

Based on a review of relevant literature and interviews with stakeholders with expertise in LMIC HCS, there is widespread consensus that LMIC HCS can be ethically acceptable if they have a sound scientific rationale and burdens to participants (and third parties) are minimised (although the appropriate weightings of particular burdens and the optimum strategies for minimising them may be contentious). While the majority of the stakeholders we interviewed were actively involved in LMIC HCS, scientists and ethicists not involved in such studies also agreed that they could be ethically acceptable—i.e., that infecting research participants with pathogens is sometimes justifiable, including where such participants are recruited from LMIC populations. Where a research question with particularly important implications for public health can only be feasibly (and/or efficiently) addressed by HCS that recruits from an LMIC population, there may be particularly strong ethical grounds for LMIC HCS. In addition, capacity building associated with HCS may lead to other benefits, including improvements in local scientific research and ethical review.

Nevertheless, there was widespread agreement that HCS can be particularly burdensome for participants, and that they sometimes involve risks to third parties. Given the burdens to participants, the need for particularly stringent informed consent processes (e.g., involving tests of understanding) is widely recognised, and such processes are already established practice in LMIC HCS. Likewise, payment of HCS participants is widely accepted; even stakeholders from Latin America, where payment is not the norm, did not think that payment was unacceptable. However, the appropriate model of payment remains contentious, and may vary in different cultural and economic settings. Given the potential for the imposition of excessive burdens (including third-party risk in particular) to undermine public trust, the need to gauge and maintain public acceptance of HCS designs is widely seen as an important reason for robust community engagement and/or social science research to occur in parallel with LMIC HCS.

There was also consensus that HCS researchers must not only be scientifically well informed, but also exceptionally careful in the conduct of HCS, particularly regarding the safety of participants. As one interviewee noted:

> No matter how careful we are in our regulations or ethical frameworks, ultimately conducting a human challenge study in the right way will come down to a conscientious, compassionate, careful investigator ... perhaps even more fundamentally than other types of research because there's such an intensive component of involvement between the researcher and the subjects, because there is this fact that the researcher is intentionally infecting people. [Ethicist, North America]

6.3   Controversies and Unresolved Issues

Our review and qualitative interviews also identified a number of controversial and/or unresolved issues in need of further empirical data and/or ethical analysis. Broadly, these relate to (i) burdens and benefits, (ii) participant selection and payment, and (iii) issues of governance.

6.3.1   Burdens and Benefits

Regarding the burdens and benefits of (LMIC) HCS, further work will be needed to (i) explore how requirements to share the benefits of research should apply to LMIC HCS (e.g., those that explore the natural history of disease and thus lead to few near-term benefits as compared with those testing a new intervention), (ii) clarify what, if any, should be considered the upper limit of risk (and/or other burdens) to which HCS participants should be exposed, (ii) determine whether a small probability of irreversible or long-term harm is an acceptable risk of HCS participation and/or how such risks should be weighed against benefits, (iii) determine how small third-party risks should be weighted in the context of background risks of infection in the community (e.g., in endemic settings), (iv) clarify the degree to which limits to risk (including third-party risk) should depend on implications for community trust and/or acceptance, (v) determine the conditions under which exposing HIC populations to the burdens of HCS with pathogens endemic only in LMICs can be ethically justified. In addition, the design and review of HCS (in LMICs and elsewhere) could be improved by more systematic risk-benefit assessment, including comparisons to alternative study designs (e.g., field trials). Such risk-benefit assessments would ideally include methods for (i) assessing controversial issues (such as those listed above), (ii) dealing with situations of uncertainty (e.g., first-in-human HCS), and (iii) making ethically acceptable trade-offs between risks to participants and public health benefits (e.g., using low risk strains in HCS might reduce risks to participants but also compromise generalisability to wild-type infection).

6.3.2   Participant Selection and Payment

Current controversies regarding participant selection and payment surround questions related to (i) the appropriate models of payment for highly burdensome HCS in different settings, (ii) whether, or when, less educated individuals should be excluded from HCS, (iii) the generalisability of adult HCS results to children (e.g., under what conditions, if any, would HCS in HIC adults be more generalisable to at-risk children in LMICs than HCS in LMIC adults?), and (iv) the conditions under which, if any, it would be acceptable to involve children in HCS. One area of consensus related to (iv) was that significant community engagement should be conducted and/or international consensus should be sought before the further planning of HCS in children.

6.3.3 Governance

The ideal model(s) of ethical and regulatory governance of HCS are yet to be determined. In particular, further work is needed to clarify (i) the appropriate model of regulatory governance of challenge strains (including in LMICs) and whether this can be standardised (to a greater degree) at an international level, (ii) the role(s) of HCS in regulatory pathways for the development of new interventions, (iii) the conditions under which HCS protocols should be reviewed by standard ethics committees and/or specialised committees, (iv) the (potential) need for a specialised ethical framework and/or principles/guidelines for HCS in general and/or HCS in LMICs—and the specific content thereof.

6.4 Future Directions

Challenge studies are a growing area of research that has the potential to advance science and lead to improvements in public health, especially in LMICs. They involve ethically sensitive research practices, however, and raise numerous controversial issues. Further work is needed by biological scientists, social scientists, community engagement experts, and bioethicists in order to establish norms of best practice for HCS that ensure the safety of participants and promote public trust and acceptance of this type of research so that its potential benefits can be realised in the long term. Further research and/or related activities regarding ethical and regulatory issues related to (LMIC) HCS could include (i) more detailed analyses of the controversial and/or unresolved issues identified in this report, (ii) broader surveys of other stakeholders (including HCS participants and members of the general public as well as policymakers in different settings), (iii) workshops with policymakers and regulatory representatives (including in endemic regions), (iv) multidisciplinary collaborations regarding HCS study design, (v) the development and refinement of risk-benefit assessment tools to compare HCS designs with one another and compare HCS with alternative designs from both ethical and scientific points of view, (vi) further research capacity building in LMICs (including the strengthening of existing ethics review mechanisms for HCS), (vii) a more extensive review of international regulatory requirements and laws regarding intentional infection, (viii) education and awareness-raising regarding HCS (including their scientific importance, ethical issues, and sharing of insights from past HCS) with stakeholders at institutions where future HCS are being considered or might be appropriate or called for, and more general public community engagement.

References

Hodgson, S.H., E. Juma, A. Salim, C. Magiri, D. Njenga, S. Molyneux, P. Njuguna, K. Awuondo, B. Lowe, and P.F. Billingsley. 2015. Lessons learnt from the first controlled human malaria infection study conducted in Nairobi, Kenya. *Malaria Journal* 14 (1): 182.

Njue, M., F. Kombe, S. Mwalukore, S. Molyneux, and V. Marsh. 2014. What are fair study benefits in international health research? Consulting community members in Kenya. *PLoS ONE* 9 (12): e113112.

Njue, M., P. Njuguna, M.C. Kapulu, G. Sanga, P. Bejon, V. Marsh, S. Molyneux, and D. Kamuya. 2018. Ethical considerations in controlled human malaria infection studies in low resource settings: Experiences and perceptions of study participants in a malaria challenge study in Kenya. *Wellcome Open Research* 3.